Falls Aren't Funny

America's Multi-Billion-Dollar Slip-and-Fall Crisis

RUSSELL J. KENDZIOR

GOVERNMENT INSTITUTES
An imprint of
THE SCARECROW PRESS, INC.
Lanham • Toronto • Plymouth, UK
2010

Government Institutes

Published by Government Institutes
An imprint of The Scarecrow Press, Inc.
A wholly owned subsidiary of The Rowman & Littlefield Publishing Group, Inc.
4501 Forbes Boulevard, Suite 200, Lanham, Maryland 20706
http://www.govinstpress.com

Estover Road, Plymouth PL6 7PY, United Kingdom

British Library Cataloguing in Publication Information Available

Library of Congress Cataloging-in-Publication Data

Kendzior, Russell J.
  Falls aren't funny : America's multi-billion-dollar slip-and-fall crisis / Russell J. Kendzior.
    p. ; cm.
  Includes bibliographical references and index.
  ISBN 978-0-86587-016-1 (cloth : alk. paper) — ISBN 978-1-60590-696-6 (pbk. : alk. paper)
  1. Falls (Accidents)—United States. I. Title.
  [DNLM: 1. Accidental Falls—prevention & control—United States. 2. Accident Prevention—methods—United States. 3. Accidental Falls—economics—United States. 4. Human Engineering—methods—United States. 5. Liability, Legal—United States. WA 288 K33f 2010]
  RD93.K46 2010
  613.6—dc22
                                                                          2009051556

∞ ™ The paper used in this publication meets the minimum requirements of American National Standard for Information Sciences—Permanence of Paper for Printed Library Materials, ANSI/NISO Z39.48-1992.

Printed in the United States of America

To all those who have been seriously injured or lost their lives from injuries sustained as a result of a slip, trip, and fall.

To Lisa, Annie, and Matthew
for their never-ending patience, love, and support.

Special thanks to Laura Cooper and Jackie Abbott
for their tireless effort in helping me complete this project.

In Remembrance of William "Bill" English (1934–2010)

"True soldiers do not fight
because they hate those in front of them
but rather because they love those
who are behind them."
—Unknown

Contents

Preface

So exactly how many people have to be injured before something is done? This is the common question that has been asked by safety professionals for decades, the cry that has led to the formation of such government agencies such as the Occupational Safety and Health Administration, the Mine Safety and Health Administration, and the National Institute for Occupational Safety and Health. Surprisingly, the reply from the government's safety agencies has, for the most part, been that of silence. With an annual taxpayer cost of $30 billion and growing, you would think that falls would be a national priority. Why have the voices of industry, trade organizations, and trade unions been silent when it comes to slip-and-fall prevention? Why hasn't the National Safety Council, the American Association of Retired Persons, and other membership-based organizations whose paying members are most likely to be affected by slips and falls responded? Where are the nation's political leaders, and why have they allowed this problem to become what the Centers for Disease Control now defines as a national epidemic?

Has our country lost its soul, its compassion, its greatness? Why is it that many of our country's largest businesses have consciously chosen not to develop slip-and-fall prevention programs but rather pay the outrageous financial costs and then turn around and blame the trial attorneys? Why is it that the media have led us to believe that most slip-and-fall claims are bogus and that the victims are scam artists seeking to defraud business owners and their insurance companies? Why is it that the insurance industry continues to pay out billions of dollars in injury claims and legal fees and, in turn, pass the costs along to their policyholders who, in turn, pass the costs along to the consumer? The costs of slips and falls are paid by all of us. One way or another, we are all victims.

Why is it that most people laugh when the see someone slip and fall—that home videos of people slipping and falling are some of the most popular YouTube downloads and helped make *America's Funniest Home Videos* one of the most-watched television programs in America?

This book attempts to answer these and other questions in a fair and comprehensive way to explain the seemingly complex relationships associated with slips and falls. This book should be the wake-up call to those government agencies, trade organizations, business owners, and insurers that the slip-and-fall epidemic is real and manageable. It just takes commitment! The idea that the financial costs associated with slips and falls is simply a cost of doing business can no longer be accepted. Those who have adhered to this failed business principle are dinosaurs who will either survive by adopting a proactive approach to safety or become extinct. This book is written with the philosophy that safety is a profit center in which everyone benefits.

I

THE PROBLEM

1

Safety Culture

When we think of a slip-and-fall accident, we often remember the old Laurel and Hardy or the Keystone Cops silent movies depicting people slipping and falling for a laugh. Slips and falls have been a cornerstone of slapstick comedy for decades, and much of this has permeated into the psyche of the general public. Until our society changes its widely held view that slips and falls are funny and begins taking the problem seriously, it is unlikely that we will see any reduction in accident claims. Even the Department of Labor under the Occupational Safety and Health Administration (OSHA) depicts accidental injuries as humorous. And these are the people in charge of protecting worker safety.

In order for a subject to be addressed, it cannot be seen as a joke. Let history be our guide. During the 1920s, the women's suffrage movement was often the target of humor, and the early feminist movement was seen as a group of radical kooks. This idea was repeated in the 1960s during the civil rights movement. As more Americans became engaged in the dialogue, both of these civil rights movements began to be taken seriously, were elevated into a "national debate," and eventually led to legislative change.

The injuries that arise as the result of a slip and fall are serious and real. Depicting slips and falls as a joke only serves to perpetuate the stereotype that such events are not real and are not to be taken seriously. Such a stereotype serves no one and leads only to the growing number of serious and unnecessary accidents.

This is also true when it comes to safety training. I am appalled by the number of safety posters, signs, and videos that poke fun at accidents. Illustrations or cartoons depicting people falling off a ladder or down a flight of stairs are actually counterproductive to building a safety message. Using such images as a means to train individuals will often lead to a culture where safety and those who promote safety are not taken seriously. I cannot tell you how often I have heard safety professionals complain that their voices are not heard within their company and that their colleagues often do not

3

take them seriously. If you want to be taken seriously, you must first work to have the subject of safety taken seriously. Stop using cartoons or "funny" safety training videos. Whether you know it or not, the tools you have been using have not served you or your mission.

AMERICA'S SAFETY CULTURE

This book could have been written about any type of accident, but it isn't. It is written about falls, specifically slips and falls—the type of accident that each and every person has experienced many times in their lives, which is why, in part, most people don't take them seriously. They're too common. From the time we took our first steps, we all have personal experience with losing our balance and falling. However, as we age, most of us pay little attention to the risks of falling, that is, until it happens to us or a loved one. Although we are the most safety-conscious nation in the world, we often fail to understand the real benefits to preventing accidents. All too often, we quantify success in terms of dollars rather than the human impact that accidents have in our society. The real measuring stick is that of the pain and suffering that is experienced from such accidents, especially those that could have been prevented. Saving money is one way of measuring success, but counting the number of prevented injuries and lives saved is even better.

As a nation, Americans believe in the importance of public safety. We inspect bridges, roads, and aircraft on a regular basis to ensure that the public remains safe. But few even think to inspect the walkways each of us walk on every day. According to the American Podiatric Medical Association, the average person takes approximately 8,000 to 10,000 steps every day, each of which represents a potential slip or trip and fall.

> In a signed memorandum of understanding with the National Floor Safety Institute, the Centers for Disease Control (CDC) describe the current rate of growth of elder falls as "a national health crisis."

> Ac'ci-dent (n). 1. Unexpected happening 2. Mishap 3. Chance — ac'ci-den'tal

THE SLIP-AND-FALL EPIDEMIC

In his book *The Tipping Point*, Malcolm Gladwell examines the concept of "social epidemics" and points out that life is a series of epidemics, each of which changes the way we view the world. These epidemics are illustrated by questions like "Why do teens smoke in greater numbers when every single person in the country knows that cigarettes kill?" or "Why did crime drop so dramatically in New York City in the mid-1990s?" One could add "Why are so many people being injured as the result of slips and falls?"

Because of medical advances, better diets, and exercise, people are living longer and fuller lives. In fact, one of the fastest-growing age-groups is centenarians—those who are 100 years old or older. More and more of us are living well beyond the age of 75 and, because of social changes over the past generation, are spending more time outside the house, traveling, shopping, or engaging in recreational activities. Age is a virus that all of us will contract and that brings with it the epidemic of fall-related injuries. As we age, we are more likely to fall and in turn face serious injury or death. By 2020, the U.S. Census Bureau estimates that the number of retirement-age individuals will reach 77 million, nearly double today's number. Because falls are the most likely cause of accidental injury affecting the elderly, as this age-group grows, so will the number of elderly falls.

Aside from the financial impact associated with falls, there is a social impact as well. Falls are the leading cause of nursing home admissions in the United States, and such injuries affect more than just the victim. Each and every person who is connected to the victim pays a price. For every parent who is seriously injured as the result of a fall, a son or daughter is affected. How many of us and our aging parents care for a grandparent who has been crippled by a fall? And who will care for the baby boomers when this epidemic reaches them? As I write this, the United States is debating a national heath care plan. Most Americans agree that health care needs to be addressed, but they disagree with the government's "single-payer" plan. Absent in the debate is the notion of accident prevention as a cost savings. Why? Because it is too simple. After all, government is about complexity rather than simplicity.

If you don't believe that injuries as a result of falls are a serious problem in need of immediate action, just ask someone over the age of 75. Most will tell you that falling is one of their biggest fears and a leading reason for seeking a home security system. It's not the burglar they fear the most but rather falling and not being able to get off the floor to call for help. And for many, a serious fall makes them dependent on others.

WOMEN ARE MORE LIKELY TO FALL THAN MEN
According to the Consumer Product Safety Commission, 60% of all falls are experienced by women, most of which occur in the home. Of the falls that take place outside the home, most occur in a retail store. Why is this? Many believe that women's footwear (that is, high heels) is to blame, and this is partly right.

However, the answer is much simpler than that. The fact is that women spend more time shopping than men do, and because many of today's retailers are building bigger stores, the average shopper takes more steps than if they were in a smaller store. Twenty years ago, an average grocery store was around 20,000 to 30,000 square feet in size, a dwarf by today's standards. It is not unusual for today's "big-box" retail stores to top 100,000 to 150,000 square feet or more, and each additional foot of retail space requires an additional footstep.

The likelihood that one step will result in a slip is directly proportional to the number of steps an individual takes.

THE CORPORATE PHILOSOPHY OF SAFETY

Most corporations are led by a senior management team whose members rose through the ranks of the marketing, accounting, or operations departments. I cannot think of a single Fortune 500 company whose chief executive officer (CEO) came up though the safety department. Because of this, most corporate leaders see the safety department as a necessary evil—something that they have to do to prevent being the target of an OSHA inspection or lawsuit. Just saying that your company has a safety manager is enough for many companies. Many safety professionals will tell you that their single biggest problem is getting their companies' management to "buy in" to improving safety. Because safety managers rarely have a budget, they have to "sell their boss" on investing in improved safety measures. Losses due to slips and falls are generally seen as a cost of doing business, a repeating line-item expense.

I can't tell you how often I have heard the comment "Sure we have accidents. That's why we have insurance." I recall a conversation I once had with a business colleague whose company sold floor-cleaning chemicals to the grocery store industry. His company was launching a new "high-traction" floor finish line designed to prevent slips and falls. On meeting with an executive from a large grocery store chain, he asked, "So exactly how big of a problem are slips and falls for your grocery chain?" The executive replied that slips and falls are their most common type of guest and employee accident. My friend then asked, "So what is your prevention strategy?" And the executive answered, "What prevention strategy? We simply raise the cost of bananas by two cents per pound and blame the increase on the trial lawyers." Welcome to corporate America, where safety is talked about but rarely invested in.

I have been the focus of many media interviews, and it never ceases to amaze me how often I learn from reporters who are interviewing me that they can't seem to get grocers to comment when they are doing a story on slips and falls. For some reason, grocers are not up to talking about their problem publicly. Why is this? Is it fear of litigation? Fear that if the facts were to become public, they are somehow more likely to be sued? A good point, especially in our litigious environment. Or should one assume that many grocers are simply unaware that slips and falls are a problem for their company? Unlikely but plausible. The response I hear the most as I have discussed this issue with the grocers I have worked with is that they would rather not comment because of the risk of being misquoted or having their statement taken out of context. Remaining silent is actually a legal strategy that helps industries, like the grocery store industry (which has a disproportionately high rate of slips and falls), to defend themselves when sued. After all, you can't use what someone doesn't say against them.

Either way, slips and falls are a serious problem for the grocery store industry and will continue to get worse until action is taken. Crisis management is how most companies respond to problems. Until slip-and-fall costs are elevated to the crisis level, management will not address the problem.

I remember a meeting I had with a large home improvement retailer. While I was waiting in the main lobby of their corporate headquarters, I picked up a copy of their annual report. One of the first pages I came across was a full-page photograph of several of their employees wearing the trademarked company apparel in an aisle of one of their stores. The caption above the image read, "Safety Is #1." However, when I reviewed the annual report more closely, I could not find exactly how much the company actually spent on safety. Although the company claimed safety to be "#1," there was no line-item expenditure for safety. This was later confirmed in my meeting with the company's managers. My conclusion was that the appearance that safety is #1 is actually #1.

Safety Managers—The First and Last Line of Defense

A March 2008 National Safety Council reader poll asked the question, "Does your CEO 'get it' about the importance of safety and health?" Sixty-two percent of those polled said yes, and the remaining 38% said no. This comes as no real surprise. For many employers, safety is something they talk about at "safety meetings" or leave to the "safety committee" but rarely is such talk elevated to the level of the CEO. After all, how many of the nation's Fortune 500 companies are led by someone who rose from the ranks of safety? Zero.

In 2003, a *Safety + Health* magazine report estimated the annual salary of safety professionals by industry and found a 58% difference in salaries for safety professionals ("Safety Professionals" 2003). The lowest-paid sector was that of agriculture, and the highest paid were employed in finance, insurance, and real estate. Surprising was that those safety professionals employed in agriculture were most likely to be protecting people, while those in finance, insurance, and real estate were most likely to be protecting assets. Nonetheless, those safety professionals charged with protecting corporate assets were paid, on average, 26% more than those who protected people.

WHEN IT COMES TO SAFETY, YOU GET WHAT YOU PAY FOR

Nation's Restaurant News, Sept. 6, 2001 If companies fail to allocate sufficient resources to safety projects and programs, they shouldn't expect to see any improvements in employee safety or any decline in workers' compensation claims, a safety and health expert emphasized at a recent conference.

Samuel J. Gualardo, a professor of occupational safety and health at Indiana University of Pennsylvania, said during a session of the annual American Society of Safety Engineers Professional Development Conference that unless supervisor and employee training and education is augmented with safety equipment and training as well as money, companies will not be able to make much progress. Gualardo pointed out that there is often too little contact between safety professionals and line managers, which results in risk managers doing what line managers should be doing. That approach means that line managers are not involved in the safety process, that they're too far away from it.

"The most important reason line managers do not feel accountable or responsible for safety performance is because it is not driven by senior management," Gualardo said. "When commitment by senior managers is not visibly demonstrated and expectations are not established or regularly communicated from the top, poor safety performance is tolerated and perceived as acceptable. If I were a line manager, I would not accept responsibility for safety either, if it were not being driven at the senior management level."

The Concept of Risk Management

We all have experience with the misinterpretation of news stories, even those that are seemingly simple. For example, one day we read a news headline outlining a new scientific study that claims that eating too much red meat can lead to an elevated risk of cancer, only to be followed up a few months later with a conflicting report (produced by an organization funded through the meat industry) that concludes the exact opposite. Who do you believe, or, more important, whom do you trust? Instinctively, we believe that there may be risk in any form of excess; that is, the more you eat red

Table 1.1. Lifetime Odds of Accidental Death

Type of Accident or Manner of Injury	Deaths	One-Year Odds	Lifetime Odds
All external causes of mortality, V01–Y89, *U01, *U03	164,112	1,755	23
Deaths due to unintentional (accidental) injuries,			
V01–X59, Y85–Y86	106,742	2,698	35
Transport accidents, V01–V99, Y85	48,366	5,953	77
Pedestrian, V01–V09	6,091	47,273	612
Car occupant, V40–V49	16,337	17,625	228
Drowning, V90, V92	413	697,194	9,019
Falls, W00–W19	16,257	17,712	229
Fall on same level from slipping, tripping,			
and stumbling, W01	646	445,729	5,766
Firearms discharge, W32–W34	762	377,876	4,888
Fireworks discharge, W39	5	57,588,244	744,997
Accidental drowning and submersion, W65–W74	3,447	83,534	1,081
Exposure to electric current, radiation, temperature,			
and pressure, W85–W99	454	634,232	8,205
Exposure to smoke, fire and flames, X00–X09	3,159	91,149	1,179
Contact with venomous spiders, X21	10	28,794,122	372,498
Lightning, X33	66	4,362,746	56,439
Earthquake and other earth movements,			
X34–X36	31	9,288,426	120,161
Cataclysmic storm, X37	63	4,570,496	59,127
Flood, X38	9	31,993,469	413,887
Assault, X85–Y09, Y87.1, *U01	17,638	16,325	211
Assault by firearm, X93–X95	11,829	24,342	315

meat, the greater the risk it has of affecting your health. But just how much is too much? How much meat can a person eat and still be considered low risk? We often fail to understand that everything we do, from eating, smoking, traveling, or even walking, involves risk. So when the average person is asked, "So what is the risk of contracting breast cancer verses that of a slip and fall?," most folks generally get it wrong.

ValuJet airlines was a fast-growing phenomena from 1994 until the deadly Everglades crash in May 1996 after which the airline was immediately grounded for two months. The crash called into question the safety of all low-cost carriers. Many industry experts didn't expect the airline to survive, and in fact, in the fall of 1997, it ceased operations. The management of the company did not expect that a seemingly small breach in a safety procedure would bring down an aircraft, let alone an airline, but it did.

Small oxygen containers, assumed to be empty, were loaded into the cargo area of the aircraft for refilling. It is believed that one or more of the containers was not empty but full. As the plane took off and gained altitude, the oxygen containers exploded under the growing pressure and ignited a fire in the cargo area that eventually resulted in the aircraft's fatal crash. In the aftermath, we saw the typical finger-pointing game between the company and the contractor who was hired to load cargo onto ValuJet's aircraft. You see, cost cutting was the rave of the airline industry during this time, and safety was discounted. Both the airline and the contractor failed in their individual responsibilities to protect the passengers of flight 592. It was argued that had smoke detectors been installed in the cargo area, the pilot may have had time to land the aircraft. Most experts agreed that warning of a cataclysmic event, such as a raging fire in the cargo hull, would provide but a few minutes of time—not enough to save the plane and its passengers.

This was the first and last serious safety event that the airline experienced. When it was all said and done, ValuJet went out of business and soon reentered the industry under the name AirTran. So for all you safety professionals who have to sell your managers on investing in safety and they ask, "Why should we spend money on safety when we have never had such an accident take place," you may want to remind them of ValuJet flight 592.

Unintentional injury death rates rise as we age. Statistically speaking, if we all live long enough, we will eventually die from injuries sustained as the result of a fall.

There are countless illustrations of the confusion most people have regarding risk. Using the comparative example of breast cancer to a slip and fall, we find many similarities. Most would assume that the single greatest risk factor for both breast cancer and an accidental fall is age—the older the woman, the greater the risk, right? However, when a 2007 survey by Oxford University researchers asked British women when a woman is most likely to get breast cancer, more than half said that "age doesn't matter." One in five thought that the risk is highest when a woman "is in her 50s," 9.3% said the risk is highest "in her 40s," and 1.3% said "in her 70s." A grand total of 0.7% of women chose the correct answer: "80 and older." Breast cancer has been a major public concern and topic of discussion since at least the early 1990s, yet the survey revealed that the vast majority of women still know nothing about the most important risk factor. How is that possible?

FEARMONGERING AND THE PSYCHE OF FEAR

When it comes to safety, we are all prompted to take notice out of fear—fear that unless we do something to reduce risk, we could be the next victim. The biggest marketer of fear is the media. More so than anyone, the media know the power of fear and use it as a sales tool. Fear sells newspapers, magazines, and books and gets us to turn our televisions on. Some believe that if it weren't for fear, there would be no need for the media. Take, for example, the alleged 2007–2008 "bird flu" crisis. We were bombarded with news about the impending pandemic. This microscopic killer disease was on our nation's doorstep, and we were all sitting ducks (no pun intended), sure not only to contract the disease but ultimately to die from it.

Organizations like the CDC, the Department of Defense, and the Department of Homeland Security were mobilized and placed on high alert. We were led, by the top media reports of the day, to believe that infection was imminent and that we were on the threshold of an international health crisis. This perfect storm was created by a network or institution of fear that was transformed into misinformation and subsequently fueled by the media. The problem was that, in order for the avionic flu to infect people, it must first make a gigantic leap from birds to humans, and this never happened. Without the migration of the virus from birds to humans, there will be no human infection. No infection means no crisis. No crisis means no news story. You get my point.

The media know the value of fear and use it as a tool to drive sales. The media are in the business of profit, and crowding in the information marketplace means that the competition for eyes and ears is steadily intensifying. With hundreds of cable news channels struggling for their piece of the news market, each is looking for that one nugget that will differentiate it from the competition. "A story you can't afford to miss!" is an excellent way to get someone's attention.

The real threat that has stemmed from the avionic flu debacle is that of the H1N1, or "swine flu," virus, which may soon rise to the level of a real pandemic. Because the avionic flu scare never materialized, many people may disregard the threats posed by future outbreaks of the flu, and that would be newsworthy.

So exactly why are we so easily consumed by fear? In part, it is who and what we are. We have evolved thousands of years with the basic instinct for self-preservation. It's simply ingrained in our brains. Another reason is what psychologists call "confirmation bias." Once a belief is in place, we screen what we see and hear in a biased way to ensure that our beliefs are "proven" correct. This alignment of values is also linked to what psychologists call "group polarization," whereby people who share beliefs group together, in turn serving as kind of a fraternity of ideas in which each member is convinced that his or her group's beliefs are right.

Psychologists discovered that every human brain has not one but two systems of thought or minds, system 1 and system 2. We know them better as feeling and reasoning.

System 2, reasoning, works slowly. Reasoning takes time and examines evidence. It calculates, considers, and rationalizes. It responds to questions such as "Will I be the victim of an accident?" with answers such as "Accidents happen to other people and not me because I'm careful." When reason makes a decision, it's easy to put into words and explain.

System 1, feeling, is entirely different from reason. Unlike reason, it works without our conscious awareness and is as fast as lightning. Feeling is the source of the snap judgments that we experience that begin as a hunch, as an intuition, or as emotions like anxiety, worry, or fear. A decision that comes from feeling is hard or even impossible to explain in words. You don't know why you feel the way you do—you just do. The problem is that the system mind doesn't work well in the modern, fast-paced, and information-intensive world we live in.

So what makes the evening news? What is it that we as safety-conscious Americans tune in to? For most, it's the rare, the unusual, the crisis of the day—who was murdered, the terrorism update, not to mention the seasonal hurricane and unpredicted plane crash. What doesn't make the news is the routine, the typical, and the expected—the disease that kills one person at a time and doesn't lend itself to strong emotional images, such as diabetes, asthma, heart disease, and, of course, accidental falls. The only time these subjects make the news is when a famous person or celebrity dies as the result of one of them.

In the late 1970s, researchers Paul Slovic and Sarah Lichtenstein showed the differences between perception and reality. Most people said that accidents and disease took lives equally even though disease inflicts about 17 times more deaths than accidents. People also estimated that car crashes kill 350 times more people than diabetes. In fact, crashes kill only 1.5 times as many. How could our perceptions be so skewed? We see high-speed car crashes all the time on the news, but only family and friends will hear of a life lost to diabetes, cancer, or a fall. The answer is that while our eyes are watching, our head sleeps.

ERRONEOUS HEADLINES

Experts don't know what constitutes a safe level of trans fat in the diet, so no "daily value" information need be included as it is for other nutrients. But, in general, given the role of trans fats in heart disease, the less of them in the diet, the better, says Tommy Thompson: "Heart disease is the number one killer of both men and women in America. In 2002, heart disease had a negative economic impact of $214 billion—that's $214 billion—including $115 billion in direct medical costs. Bad fats like trans fats and saturated fats contribute heavily to obesity as well."

"Several years ago, the message was, 'oh no, caffeine is bad, bad, bad,'" says registered dietitian Lona Sandon. "Now more research has come out and that doesn't seem to be the case."

Millions of people who thought they had healthy blood pressure are about to get a surprise: the government says that levels once considered normal or borderline actually signal "prehypertension," and those people must take care to stave off full-blown high blood pressure.

Numbers can also be confusing for many people. We can't picture in our minds 100, 1,000, or 100,000 lives, so we don't *feel* 100, 1,000, or 100,000 lives. Humans are better at feeling proportions and can better understand percentages rather than size. Fifty, 75, or 99% of something means more than the something it represents. Back in the 1990s, the chairman of the ASTM Committee D-21 suggested that the likelihood of an accidental slip on a surface coated with a floor finish was one in a billion. That is, it would take a billion steps in order for one individual's step to result in a slip. To most people, this statistic would imply that it is virtually impossible for people to slip on a polished floor, but quite the contrary. What is grossly misunderstood is the fact that according to the American Podiatric Medical Association, the average person takes between 8,000 and 10,000 steps each day. Multiply this number by the number of Americans, and you will find that the seemingly low probability of slipping on a polished floor predicts more than 1 million slip-and-fall events each year, about the number that the Consumer Product Safety Commission reported being treated at the nation's emergency rooms.

Our feelings are more often influenced by experience and culture, and we live in a culture of fear. Fear of what could happen is more easily understood than what has happened in the past or what is currently happening.

In his book *The Science of Fear*, author Daniel Gardner discusses the 2005 survey conducted by Dan Kahan of the Yale Law School along with Paul Slovic and others in which they surveyed 1,800 Americans to gauge their perception of risk.

Past surveys had shown that nonwhites viewed risks higher than whites and that women more than men believed that risks were more serious. When put together, this is called the white-male effect. In general, white men more than other people feel that hazards are less serious. But Kahan and Slovic's survey showed that not all white males felt this way. Only 30% of white males perceived things as less dangerous than did others. Their survey also revealed that the confident minority of white men tended to be better educated, wealthier, and more politically conservative than others. But what exactly made this group different was not their race or sex but rather their cultural ties. According to Gardner,

> What Kahan had found was a strong correlation between risk and other factors like income and education. But the strongest correlations were between risk perception and world view. If a person were, for example, a hierarchic, someone who believes people should have defined places in society and respect authority, you could quite accurately predict what he felt about various risks. Abortion? A serious risk to a woman's health. Marijuana? A dangerous drug. Climate change? Not a big threat. Guns? Not a problem in the hands of law-abiding citizens.
>
> Kahan also found that a disproportionate number of white men were hierarchs or individualists. When he adjusted the numbers to account for this, the white-male effect disappeared. So it wasn't race and gender that mattered. It was culture." (Gardner 2008)

So how does this apply to safety? Given this understanding of risk as a function of one's culture, we can build a new risk-investment model. If the management of a particular company is comprised of hierarchical, white men who view the world as low risk, it is unlikely that they will invest in safety, not because they as a group believe that safety (i.e., accident prevention) isn't important but because they believe that it is not a significant problem and therefore not worthy of significant attention or investment.

2

The Slip-and-Fall Problem

CAUSES

Research, based in part on insurance industry claims data, has revealed five major causes responsible for almost all slips, trips, and falls. Although the actual percentages may vary from one industry to another, the following five causes have been well documented across different industry groups, environmental conditions, and geographies.

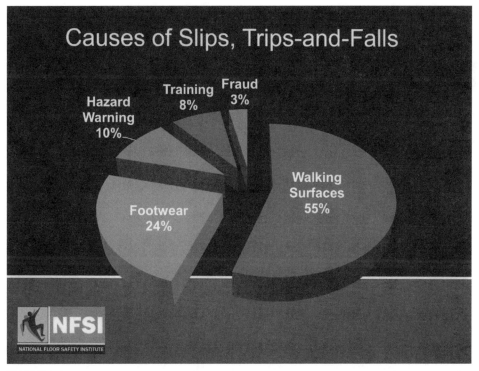

FIGURE 2.1
Causes of Slips, Trips, and Falls. From National Floor Safety Institute.

Although the walking surface is most likely to be identified as the primary cause of a slip-and-fall accident, making up 55% of all falls, the remaining 45% are attributable to four other factors, including footwear, fraud, hazard identification, and training. Therefore, a property owner's slip-and-fall prevention strategy should focus on a comprehensive approach that addresses all causes, especially their floors.

Only 3% of slip, trip, and fall claims are fraudulent.

CONTRIBUTING FACTORS TO SLIPS, TRIPS, AND FALLS

When examining the factors contributing to slips and falls, we find that they can be divided into three distinct categories: physiological, social/emotional, and environmental. In my experience as an expert witness, I have found that most slip-and-fall incidents take place as a result of a combination of factors most of which take root in some type of walkway hazard.

Physiological

Examining the first contributing category, the physiological, it is important to recognize how each of us differs in physiology and how this directly affects our ability to walk safely. Walking is nothing short of a constant balancing act; from the day we took our first steps, we have known what it is like to fall. However, as humans grow older, they tend to take for granted the importance of good balance, only to be reminded in old age. Lack of balance is the leading physiological factor contributing to slips and falls. It is a function of several physiological conditions, including the inner ear, visual stimuli, sensory motor input, skeletal structure, and, perhaps most important, the brain itself. Together, these physiological factors create our individual consciousness, which in turn provides us an awareness of our personal safety. If for any reason we misinterpret environmental conditions, an accident becomes more likely.

PHYSIOLOGICAL FACTORS CONTRIBUTING TO SLIPS AND FALLS
- Balance (poor hearing)
- Brain (motor impairment)
- Muscle tone (foot, thigh, calf strength)
- Skeletal size and strength
- Visual impairments
- Foot shuffle

STUDY LINKS INNER-EAR DISORDER TO FALLS

More than 35% of Americans age 40 and older suffer from vestibular dysfunction, an inner-ear disorder that can increase the risk of falls, according to a study from the Johns Hopkins University School of Medicine in Baltimore.

The study said that participants included 5,086 adults who performed balance tests and completed a questionnaire on their history of dizziness and falls. Results found that 35.4% of the population age 40 and older suffer from the condition, with its prevalence increasing significantly with age. An increased risk was also seen among people with diabetes and those with a high school education or less.

People experiencing symptoms of vestibular dysfunction had an eightfold increase in their risk of falling and an increased risk of hearing loss. Study authors called for screenings for the condition in nursing homes and assisted living facilities.

Social

The second contributing category—and a close kin to the first—includes the social/emotional factors affecting a slip-and-fall victim. Many demographics and statistics surrounding falls describe in part the social factors. Factors like the victim's age, sex, and disability play a role in defining fall data. However, rather than being mere coincidences, these factors directly contribute to falls. A good example is that of concentration or preoccupation. Each day, thousands of people use the roads and highways to travel to work. For many, the actual journey to the office is subconscious. This spatial familiarity is common in many of our daily lives and often leads to accidents. The old adage "watch where you are walking" holds true in the prevention of slips and falls. Unfortunately, for many people, this may be easier said than done.

SOCIAL FACTORS THAT CONTRIBUTE TO SLIPS AND FALLS

- Age
- Sex (hormonal variations)
- Physical disabilities
- Health (medications, drugs)
- Mental frame of mind
- Concentration/preoccupation

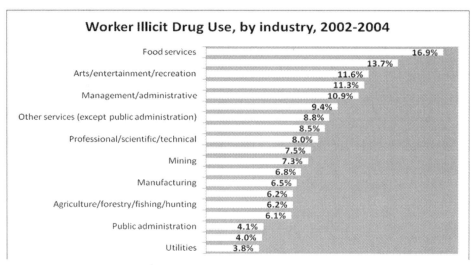

FIGURE 2.2
From National Safety Council.

One of the contributing factors of slips and falls is medication, including illicit drugs. Most drugs have side effects, such as loss of balance, dizziness, and blurred vision, all of which may increase the risk of a slip and fall. Surprisingly, the restaurant industry has one of the highest rates of illicit drug problems, and this may contribute to the rising rate of restaurant slips and falls.

TWO-THIRDS OF U.S. WORKERS DON'T CARE ABOUT THEIR WORK

If you're trying to get your workers to think more about safety, you may first want to help them care more about their jobs. It could be the key to decreasing your injury numbers.

A recent study found as many as two-thirds of today's workers are either looking for another job or are just going through the motions. The study found that even the traditional motivators, such as incentives or competitive pay, are becoming less effective in getting workers to value their jobs.

Bottom line: Safety incentives often let managers down when it comes to preventing injuries. When workers don't care, they have more accidents ("Two-Thirds of U.S. Workers Don't Care about Their Work" 2007).

Environmental

The third and final contributing category is that of environmental conditions. For many people—and certainly those who have been the victim of a slip and fall—this factor is often seen as the only "cause" of a slip and fall. This is untrue. What is true is that given the wrong set of circumstances, environmental factors are most likely to be identified as the cause of the accident. For example, assume that an elderly person who suffers from poor vision enters a business and slips and falls on a spot of water that he or she did not see. What was the cause? This type of slip and fall is more likely to be the norm than the exception, and placing responsibility exclusively on the victim or the property owner is difficult.

What is unique about the environmental category, unlike the first two, is that this category is the only one that the property owner can directly control and therefore be held liable for. Conditions such as the flooring material and maintenance, lighting, and hazards are within the control of the property owner. Also, for those workplaces that have wet, greasy, or otherwise hazardous flooring conditions as a part of the work environment, employee footwear can be seen as a controllable environmental factor. Mandating appropriate footwear is one of the most overlooked prevention strategies in the workplace today, and if implemented by the employer, it can reduce work-related slips by 25% or more.

ENVIRONMENTAL FACTORS THAT CONTRIBUTE TO SLIPS AND FALLS
- Walkway surfaces
- Improper floor maintenance procedures
- Walkway hazards
- Slope and elevations
- Walkway transitions
- Walkway hazards and improper identification
- Mats and rugs
- Footwear (type and fit)
- Lighting (visual hazards)
- Stairway design
- Distractions (displays, advertisements)

Most slip-and-fall accidents are caused by a combination of contributing factors. Rarely is there a single cause, and the business owner or risk manager examining such claims would be well served to focus on a comprehensive solution to prevention rather than a single-cause strategy. Each of the previously named contributing factors plays a role in how and why slips and falls occur.

Although property owners cannot control the physiological and social factors that affect both their employees and invited guests, they can control the environmental

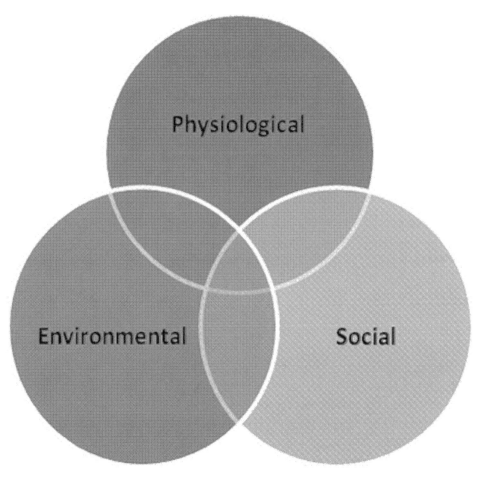

FIGURE 2.3

factors of their property. To reduce the risk of a slip-and-fall accident, it is suggested that property owners institute a strict walkway safety program that incorporates the use of high-traction floor-covering materials, floor care products, employee training in slip-and-fall-accident prevention, and frequent slip resistance audits of their walkways.

Research shows that during normal ambulation, an average person lifts his or her feet only a quarter inch above floor level.

PET OWNERS BEWARE: STUDY SHOWS 86,000 FALLS ANNUALLY ARE CAUSED BY PETS

For the first time ever, the Centers for Disease Control and Prevention conducted a study focusing on the relationship between falls and pets. The results, which are published in this week's CDC's [Centers for Disease Control] *Morbidity and Mortality Weekly Report*, reveal that cats and dogs are the cause of over 86,000 falls each year. Dogs were to blame for a vast majority of these accidents by causing owners to trip by running in front of their legs, pulling on the leash during a walk, or taking off after squirrels and other animals leaving their owners to chase after them unable to watch their step. There are also reports of tripping over dog toys (Devon Glass, http://www.injuryboard.com, March 27, 2009).

Although dogs are the primary factor in pet-related falls, cat owners are not exempt. Researchers discovered that cats caused owners to lose their balance when the cats rubbed against the owners' legs, causing them to trip. Although most of these injuries are relatively minor, for senior citizens minor falls can be devastating. Out of every 10,000 people who visit emergency departments each year for slip-and-fall injuries, only about three are related to pets. However, that number doubles for people over the age of 75.

HOW ARE THE DATA COLLECTED?

Before we start this process of understanding just how big the problem of slips, trips, and falls really is, we must first understand how data are collected and interpreted. Listed next are the leading sources of data collection as it applies to slips, trips, and falls.

1. The U.S. Bureau of Labor Statistics created the Census of Fatal Occupational Injuries (CFOI), a federal–state cooperative program using death certificates; state workers' compensation reports; coroner, medical examiner, and autopsy reports; Occupational Safety and Health Administration (OSHA) fatality reports; and news stories.
2. The National Safety Council (NSC) utilizes the CFOI program and uses state vital statistics and workers' compensation reports to make estimates for the current year.
3. The National Institute for Occupational Safety and Health (NIOSH) utilizes death certificates obtained from the states.
4. The National Floor Safety Institute merges government data into independently collected data collected through the insurance industry and self-insured property owners.

5. The Consumer Product Safety Commission (CPSC) relies on data collected from the National Electronic Injury Surveillance System (NEISS), which provides estimates of consumer product–related injuries based on a national probability sample of hospital emergency rooms.

Other sources include the following:

- State workers' compensation programs
- Private insurers' data
- Private employers' data

NATIONAL SAFETY COUNCIL STATISTICS
(*INJURY FACTS*, 2007 AND 2008 EDITIONS)
- 21,200 Americans died as a result of a fall in 2007, a 2% increase from 2005.
- In 2005, there were 6,600 public unintentional-injury deaths from falls (24% of all public unintentional-injury deaths); this was a 6% increase from 2004. This includes deaths from falls from one level to another or on the same level in public places but excludes deaths from falls in moving vehicles.
- In 2005, there were 10,400 home unintentional-injury deaths from falls (28% of all home unintentional-injury deaths); this was a 1% decrease from 2004. It includes deaths from falls from one level to another or on the same level in the home or on home premises.
- About 8.5 million visits to emergency departments were made in 2004 because of accidental falls.
- Between 1999 and 2003, the number of fall-related deaths among persons 65 and older rose 36%, while the number of deaths increased by only 15% among those under 65 years old.
- Falls were the leading cause of nonfatal injuries for all age-groups except those 15 to 24 years old. The leading cause of nonfatal injuries for this age-group was being struck by or against an object.
- In independent older community-dwelling people, about 50% of falls occur within their homes and immediate home surroundings. Most falls occur on level surfaces within commonly used rooms such as the bedroom, living room, and kitchen. Comparatively few falls occur in the bathroom, on stairs, or from ladders and stools. While a proportion of falls involve a hazard such as a loose rug or a slippery floor, many do not involve obvious environmental hazards. The remaining falls occur in public places and other people's homes. Commonly reported environmental factors involved in falls in public places in-

clude pavement cracks and misalignments, gutters, steps, construction works, uneven ground, and slippery surfaces (Lord, Sherrington, and Menz 2001).

- Unintentional injuries are the fifth-leading cause of death overall and first among persons in age-groups from 1 to 44. By single years of age, unintentional injuries are the leading cause from ages 1 to 41.

SAME-LEVEL SLIP-AND-FALL STATISTICS

Here are some quick facts related to same-level slip-and-fall statistics:

- Fall fatalities are nearly equally divided between men and women. However, more women will experience a slip-and-fall accident. According to the U.S. Bureau of Labor Statistics, falls accounted for 5% of the job-related fatalities for women compared to 11% for men.
- Falls account for over 8 million hospital emergency room visits, representing the leading cause of visits (21.3%). Slips and falls account for over 1 million visits, or 12% of total falls.

Table 2.1. Unintentional Injury Deaths in the Home, 2004

Cause	Total Deaths	Change from 2003	Death Rate*
Falls	12,800	+8%	4.4
Poisoning	9,600	−17%	3.3
Fires/flames/smoke	3,500	+17%	1.2
Choking	2,000	−9%	0.7
Mechanical suffocation	900	0%	0.3
Drowning	800	+14%	0.3
Natural heat or cold	600	+20%	0.2
Firearms	500	0%	0.2
All other	6,700	+6%	2.3

Source: National Safety Council (2006).
* Rate is number of deaths per 100,000 population.

Table 2.2. Leading Causes of Unintentional Injury Deaths for All Ages, 2003 Data

Motor vehicle	44,757
Poisoning	19,457
Falls	17,229
Choking	4,272
Fires and flames	3,369
All other unintentional injuries	20,193
Total	109,277

Source: National Safety Council (2008).

- Fractures are the most serious consequences of falls and occur in 5% of all people who fall.
- Slips and falls do not constitute a primary cause of fatal occupational injuries, but they do represent the primary cause of lost days from work.
- Slips and falls are the leading cause of workers' compensation claims and are the leading cause of occupational injury for people aged 15 to 24 years.
- According to the CPSC, floors and flooring materials contribute directly to more than 2 million fall injuries each year.
- Half of all accidental deaths in the home are caused by a fall. Most fall injuries in the home happen at ground level, not from an elevation.
- Of all fractures from falls, hip fractures are the most serious and lead to the greatest health problems and number of deaths.
- Each year in the United States, one of every three persons over the age of 65 will experience a fall, half of which are repeat fallers.
- According to the CDC, in 2005, more than 158,007 people over the age of 65 died as a result of a fall, up from 7,700 a decade earlier.
- The CDC also reports that approximately 1.8 million people over the age of 65 were treated in an emergency room as a result of a fall.
- For people aged 65 to 84 years, falls are the second-leading cause of injury-related death; for those aged 85 years or older, falls are the leading cause of injury-related death.
- Incidence of falls goes up with each decade of life.
- Of all deaths associated with falls, 60% involve people aged 75 years or older.
- Falls account for 87% of all fractures among people over the age of 65 and are the second-leading cause of spinal cord and brain injury.
- Half of all elderly adults over the age of 65 hospitalized for hip fractures cannot return home or live independently after the fracture.
- Falls represent 40% of all nursing home admissions and are the sixth-leading cause of death among people aged 70 years or older.
- Over 60% of nursing home residents will fall each year.
- According to the National Institute on Aging, every year 30% of people over the age of 65 will sustain a fall, of which 10% will result in a serious injury.
- Sixty-seven percent of fall fatalities are among people aged 75 years or older.
- People over the age of 85 are 10 to 15 times more likely to experience a hip fracture than are people aged 60 to 65 years.
- Eighty-five percent of workers' compensation claims are attributed to employees slipping on slick floors (*Industrial Safety and Occupational Health Management*, 5th ed.).
- Twenty-two percent of slip-and-fall incidents resulted in more than 31 days away from work (U.S. Bureau of Labor Statistics 2006).

- Compensation and medical costs associated with employee slip-and-fall accidents is approximately $70 billion annually (National Safety Council, *Injury Facts*, 2003 ed.).

- Occupational fatalities due to falls are approximately 600 per year, down from 1,200 since the 1970s.

- Total injuries due to falls estimated at $13 million to $14 million per year in United States. Falls are the leading cause of accidental injury, resulting in 20.8% of all emergency room visits in 1995. (Motor vehicle accidents accounted for 11.9% of such visits.)

- Disabling (temporary and permanent) occupational injuries due to falls are approximately $250,000 to $300,000 per year.

- Falls occur in virtually all manufacturing and service sectors. Fatal falls, however, are in construction, mining, and certain maintenance activities.

- According to workers' compensation statistics from the ITT-Hartford Insurance Company, falls account for 16% of all claims and 26% of all costs. This compares to 33% of costs associated with sprains and strains.

- According to the American Trucking Association, slips and falls are the leading cause of compensable injury in the trucking industry.

- Falls from elevation (approximately 40% of compensable fall cases, approximately 10% of occupational fatalities).

- Falls on the same level (approximately 60% of compensable fall cases) (Keyserling 2000).

INTERPRETATION OF FALL DATA

As the old saying goes, "Figures lie, and liars figure." I caution the reader to examine accidental fall data with care. When reading fall statistics, it is important to understand that fall data do not cover only slips and falls but rather include falls of all types. For example, the National Safety Council claims that every year, more than 300,000 individuals are victims of a fall-related injury. Furthermore, falls are the second-leading cause of accidental death for individuals aged 65 years or older. What these statistics do not reveal is how and why these individuals fell. Were they slips and falls from the same elevation? Falls downstairs? Falls from an elevation? Or maybe each simply lost his or her balance. Before using fall data, qualify the statistics. Having researched same-level slips and falls for more than 20 years, I urge the reader to scrutinize the data received to ensure that such data are interpreted properly and that progress is tracked accurately. All too often, fall-related data have little or nothing to do with falls. Also remember that, as a rule of thumb, for every recorded incident, there are 10 unrecorded incidents, meaning that the real slip, trip, and fall problem may be 10 times greater than data reveal. And when we examine fall-related events, we find that for every fatality there are 23 serious injuries and 20 moderate or unrecordable events.

Another source of misleading fall-related data often comes by way of the insurance industry. Each insurance company has a different way of coding its claims data. A simple fall can be coded by an insurer many different ways (i.e., fall downstairs, from an elevation, from the same level, and so on). Another problem with insurance industry claims data is causation. I have often seen a slip and fall listed as something else. Insurers often enter claims data into their system not by the root cause but rather by the end result. A good example is from the food industry—a worker slips as a result of a slippery floor and does not fall but catches him- or herself on a hot surface or sharp object. Although the result was a cut or burn to the hand or arm, the cause was a slippery floor.

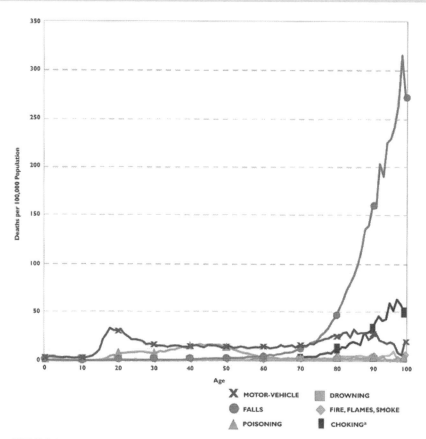

FIGURE 2.4
Unintentional-Injury Deaths per 100,000 Population by Age and Event, United States, 2002. From National Safety Council Injury Facts 2008.

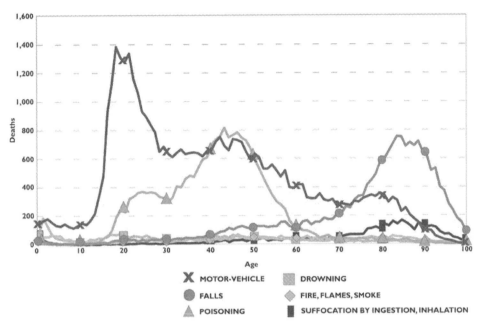

FIGURE 2.5
Unintentional-Injury Deaths by Age and Event, United States, 2002. From National Safety Council Injury Facts 2008.

Accidental deaths include deaths from falls from one level to another or on the
same level. Excluded are falls in or from transport vehicles or while boarding or alight-
ing from them. (NSC, *Injury Facts*, 2008 ed.)

Since 2005, the number of unintentional deaths have risen by 2% to 21,200, a
death rate of 7.1.

WOMEN ARE THE MOST LIKELY VICTIMS OF A SLIP, TRIP, OR FALL
The most likely victim of a slip, trip, or fall is a white female over the age of 60
with a thin body structure who has a family history of a fall. The typical female
fall victim may have an early onset of menopause, low blood pressure, and a
low-calcium diet. Finally, it is common for victims of slips, trips, and falls to be
diabetic or to have a foot disorder of some type.

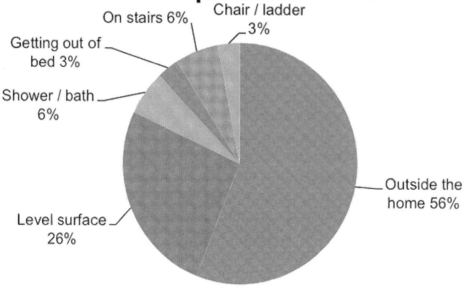

FIGURE 2.6

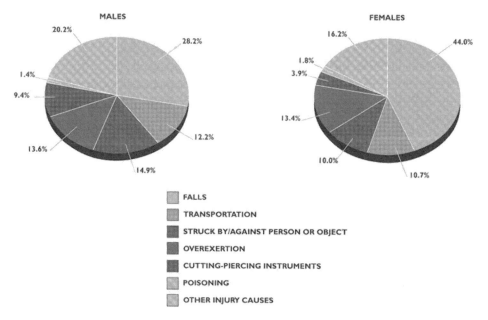

FIGURE 2.7
Leading External Causes of Injury and Poisoning Episodes by Sex, United States, 2005. NSC Injury Facts, 2008.

COSTS OF SLIP, TRIP, AND FALL INJURIES

According to the NSC, the total cost of unintentional injuries in 2006 topped $652.1 billion, which included economic costs of fatal and nonfatal unintentional injuries as well as employers costs, uninsured costs, vehicle damage costs, and fire losses.

It is estimated that when the public- and private-sector costs are combined, the total cost of unintentional slips, trips, and falls costs the nation nearly $80 billion each year and may double with the next decade.

HOW BIG IS $80 BILLION?

With $80 billion, you could pay the following:

- The salaries of 2.2 million American workers for a year
- All personal income taxes for 7.4 million Americans for a year
- Tuition for nearly 15.6 million students at America's four-year public universities for a year
- Health care costs for nearly two of every three seniors aged 65 and over for a year
- Every chief executive officer of America's 500 largest companies for the next 16 years

You could also do the following:

- Build 293,000 new homes
- Buy a new car for nearly 3 million licensed drivers (enough for every driver in New Jersey or Georgia)
- Send everyone in Florida, Colorado, and South Dakota to the movies every day for a whole year (popcorn and soda are extra!) (That's 23 million people!)
- Buy coffee and a bagel (with cream cheese) for every adult in New York City (6 million people) each morning for 20 years
- Fund America's entire space program for the next five years or launch 62 space shuttle missions
- Build 69 stealth bombers
- Send 6 million families of four to their local Major League Baseball home games for an entire season. (They get tickets, caps, programs, sodas, beer, and hotdogs—and parking, and mustard's free!)

- Plant enough trees to reforest a logged-out area the size of New England, New York, Pennsylvania, Ohio, and North Carolina—combined
- Pay for nearly half of all prescription drugs now bought by Americans each year
- Fund all cancer research in America for the next 13 years
- Buy enough oil to power every car, sport utility vehicle, and light truck for seven and a half months
- Music lovers: Buy 100 CDs a day for nearly 129,000 years before you run out of money (How long is that? About when Neanderthal man first appeared until now.)
- Eighty billion one-dollar bills placed end to end would stretch to the moon and back—about 16 times (Coalition Against Insurance Fraud 2008).

INDUSTRY-SPECIFIC DATA

Restaurant/Food Service Industry

Slip-and-fall accidents top the list of employee injuries (*Nation's Restaurant News*, September 6, 2001) and are the second-leading type of guest injury in the restaurant and food service industry. According to the National Restaurant Association, nearly 3 million employees and 1 million customers are injured in slip-and-fall accidents in restaurants each year, costing businesses an average of $150,000 per incident.

The food service industry's leading cause of employee injury, slipping and falling is also one of the leading causes of injury in the American workplace. Whether in offices, docks, factories, hospitals, schools, reception areas, or kitchens, the incidence of employees getting hurt as a result of slipping or falling is the second most prevalent cause of worker injury, according to Liberty Mutual, a leading underwriter of workers' compensation insurance and other disability-related insurance.

In a first-of-its-kind report, Liberty Mutual released a Workplace Safety Index measuring the costs and types of injuries incurred by workers on the job in the course of a year. Using data from its own claims experiences as well as data from the U.S. Bureau of Labor Statistics from 1998, the most current year with such information, Liberty Mutual reported that employers paid $4.4 billion in medical costs and workers' compensation claims to employees who injured themselves from a "fall on the same level" or a slip and fall. The Liberty Mutual index also found that the top ten types of job-site accidents account for 86% of the $38.7 billion in wage replacement and medical payments made in 1998.

Tom Leamon, vice president of Liberty Mutual and director of the Hopkinton Research Center, where the insurer experiments with and attempts to identify work-

Table 2.3. Combined Data for Slip-and-Fall Accidents: National Floor Safety Institute (NFSI), National Safety Council (NSC), Jury Verdict Research (JVR), Wausau Insurance, Liberty Mutual, U.S. Bureau of Labor Statistics, and C.N.A. Insurance

Expense
- Approximately 25,000 people a day are victims to slip-and-fall accidents. The expense of these injuries is running $3.5 million per hour, every hour of the day, every day of the year. That's over $30 billion per year (NSC).

- The average cost of a slip-related injury exceeds $12,000 (Wausau Insurance).

- The average cost to defend a slip-and-fall lawsuit is $50,000 (NFSI).

- Plaintiffs win 51% of premises liability claims (JVR).

Food Service/Hospitality
- 57% of all food service general liability insurance claims are slip-and-fall accidents (C.N.A. Insurance).

- Since 1980, personal injury lawsuits have risen industry wide (NFSI).

- The average restaurant has three to nine slip-and-fall accidents each year (NSC).

- Slips and falls are the leading cause of accidents in hotels, restaurants, and public buildings; 70% of these accidents occur on/flat level surfaces (U.S. Bureau of Labor Statistics).

Employee
- 65% of all lost workdays are due to slip-and-fall accidents. This results in 95 million lost workdays per year (Liberty Mutual).

- 22% of slip-and-fall incidents resulted in more than 31 days away from work (U.S. Bureau of Labor Statistics).

- Compensation and medical costs associated with employee slip-and-fall accidents total approximately $70 billion annually (NSC).

- The food service industry's leading cause of employee injury is slipping and falling (Liberty Mutual).

Retail/Supermarket
- The nation's $495 billion grocery store industry spends $450 million annually to defend slip-and-fall claims (NFSI).

- Slip-and-fall accidents in supermarkets are the leading cause of both employee and guest injuries (NFSI).

- Nearly 60% of all grocery store general liability insurance claims are slip-and-fall complaints (NFSI).

- For every $1.00 spent on floor care, supermarkets spend $3.00 for slip-and-fall claims (NFSI).

place hazards, said that "what's more disturbing about the high costs associated with slip-and-fall accidents is the high number of workplace deaths that result from them." "Four thousand people die from falling over on the same level on the job," Leamon said. "That's staggering. We're not talking about falling downstairs. That's one jumbo jet a month. If the FAA had to deal with those kinds of numbers, jumbo jets would never fly again."

Leamon estimates that restaurant slip-and-fall injuries affect approximately 4,000 people a year, and the National Safety Council puts it at closer to around 8,000. Leamon said that one of the ironies about slipping and falling is that many such accidents occur when employees are cleaning up the floor.

"Whether it is water or grease, fluids make floors slippery and dangerous," he said. "If you soil a floor and do the right thing and clean it up with a mop and water, you've got to get the water off the floor, too. We recommend using a squeegee. But drying the floor makes it better for traction. This is really not rocket science. Whatever goes on the floor has to be removed from the floor."

RESTAURANT EMPLOYEES ARE USUALLY YOUNG

Data from the 1990 census indicate that most youths worked in the retail trades—especially restaurants and grocery stores (National Institute for Occupational Safety and Health [NIOSH] 1997). CPSC data indicate that 87% of employed adolescents aged 16 and 17 worked part time (34 or fewer hours per week) in 1995. Data from the 1990 census showed that adolescents aged 16 and 17 worked an average of 24 hours per week for 25 weeks of the year in 1989 (NIOSH 1997). Longitudinal surveys demonstrate that brief periods of employment are common among young workers (U.S. Department of Labor 2009).

Figure 2.9 lists the most common jobs held by teens. Restaurants top the list.

Falls on the same level are the second-leading cause of injury.

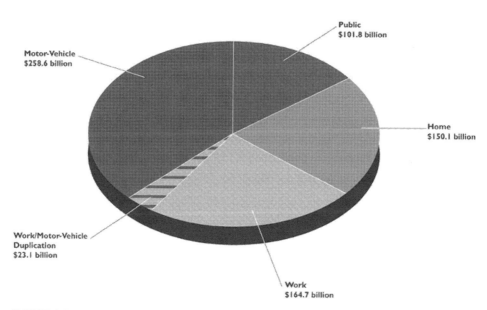

FIGURE 2.8
Cost of Unintentional Injuries by Class, United States, 2006. NSC Injury Facts, 2008.

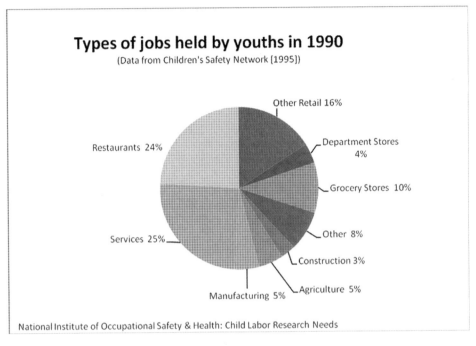

Types of jobs held by youths in 1990
(Data from Children's Safety Network [1995])

Other Retail 16%

Department Stores 4%

Grocery Stores 10%

Other 8%

Construction 3%

Agriculture 5%

Manufacturing 5%

Services 25%

Restaurants 24%

National Institute of Occupational Safety & Health: Child Labor Research Needs

FIGURE 2.9

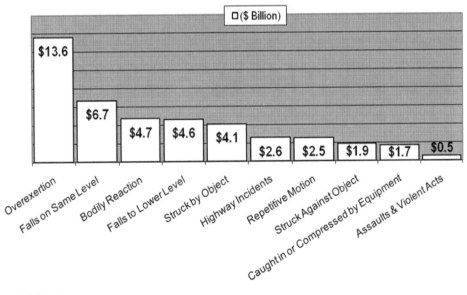
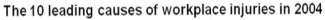

The 10 leading causes of workplace injuries in 2004

□ ($ Billion)

$13.6

$6.7

$4.7

$4.6

$4.1

$2.6

$2.5

$1.9

$1.7

$0.5

Overexertion

Falls on Same Level

Bodily Reaction

Falls to Lower Level

Struck by Object

Highway Incidents

Repetitive Motion

Struck Against Object

Caught in or Compressed by Equipment

Assaults & Violent Acts

FIGURE 2.10
From 2006 Liberty Mutual Workplace Safety Index.

TRAINING

Ask how much training the teenager will receive, and you may be surprised. NIOSH cites a survey conducted in Massachusetts that suggests that only 50% of teenagers reported receiving any training. Also troubling is the fact that teen workers tend to be unaware of their legal rights on the job more than the adults around them. Task-specific training is another concern. OSHA requires training in basic chemical exposures (29 CFR 1910.1200), blood-borne pathogens (1910.1030), and the use of fire extinguishers (1910.157[g]). There are many similar *mandatory* requirements.

Another area of interest is personal protective equipment (PPE). OSHA requires training regarding hazards that require protection for eyes, face, head, and extremities (1910.132[a]). The regulations also require training for each employee relative to the particular kind of PPE used (1910.132[f]) and mandate that the employer verify that the employee understood the training (1910.132[g]).

It may seem amazing that an employer would fail to offer any training to employees. As a parent, however, that failure is much more poignant. Imagine not training a teen how to use a bench grinder or not even offering safety glasses. Most teens can't clean their rooms at home, let alone properly clean a floor in a busy restaurant.

SUPERVISION

NIOSH found that among teenage workers, 15% to 20% either "often or always" worked with no supervision (Winn, Winn, and Biddle 2007). That is a formula for disaster—and it is not the experienced worker who will get hurt. One good example is that of proper footwear. Just about every large restaurant chain in America has a "recommended" footwear policy, but not all employees follow such a policy. The next time you go to a quick-serve restaurant, take a look over the counter and see if the employees are wearing their "required" slip-resistant shoes. Chances are, many are not—including their supervisor.

If supervisors are not committed to safety, how can we then expect the workers to be?

Wet, soiled restaurant floors are more slippery than dry, soiled floors.

On average, restaurant soils consist of the materials shown in table 2.4. Particulates are essentially clay particles that are tracked into the restaurant from outside. They become "glued" onto the floor by organic soil. The organic soils consist of triglycerides and their breakdown products. Triglycerides come from both food fats and frying oils. During frying operations, some of the triglycerides break down into fatty acids. Triglycerides and fatty acids are fairly easy to remove with most floor cleaners. Triglyceride polymers, on the other hand, represent the greatest cleaning challenge and must therefore be included in any realistic soil model. They are formed by the reaction between triglycerides and air, causing triglycerides to link up with each other, forming successively longer triglyceride polymers. The shorter polymers are sticky, and the longer polymers form a very tough film. This reaction is chemically similar to the reaction that takes place in paints that are based on linseed oil but takes place over a longer period of time. These polymers can significantly alter coefficient-of-friction behavior by virtue of their sticky and/or varnished characteristics.

Hospitality Industry

The costs of sprain and strain injuries represent almost 80% of all injury dollars spent in the hospitality industry. These injuries include lifting-related injuries, overexertions, cumulative trauma disorders, and slips and falls. Forty-two percent of employee and guest incidents were due to a slip-and-fall event (Kohr 1991), making such injuries the leading cause of both employee and guest injury. The top five most common locations for these injuries include the following:

- Bathtubs and showers
- Entrances and landings
- Ramps
- Parking lots
- Balconies

Shopping Malls/Retail

Slips, trips, and falls represent the second-leading cause of personal injury for shopping malls, following automobile crashes (Finkel and Bitzer 1998). The retail industry has the highest incidence of "same level" falls when compared with all other industries.

Table 2.4. Makeup of Restaurant Soils

Particulate soil	10%
Food fats and oils	
Fatty acids	9%
Triglycerides	71%
Triglyceride polymers	10%

Table 2.5.   Most Costly Guest Injuries and Illnesses

% Total Cost	% Total Claims	Average Cost per Injury	Accident Type
34%	41%	$20,125	Slips and falls—on wet floors, parking areas, and stairs

Source: Employers Insurance Company of Wausau (2006a). Based on a study of 1,836 major injuries costing $45,204,271.

Table 2.6.   Most Costly Guest Injuries and Illnesses

% Total Cost	% Total Claims	Average Cost per Injury	Accident Type
30%	28%	$23,674	Slips and falls on level surfaces—floors, sidewalks, or parking areas

Source: Employers Insurance Company of Wausau (2006b). Based on a study of 2,452 disabling injuries costing $53,810,104 for workers' compensation classes 9050 and 9052.

These types of claims account for approximately 17.1% of workers' compensation claims and from 1996 to 1999 represented 22.4% of the total cost to industry (Cotnam, Chang, and Courtney 2000).

While there is no centralized source for data concerning retail customer injuries, same-level customer falls are the leading source of loss for the general liability claims, representing approximately 33% of the 1999 general liability claims and 35% of the 1999 general liability losses as reported by Liberty Mutual.

Spotlight on Retail Trade
What caused your injury?
LEADING SOURCES OF INJURY OR ILLNESS 2005

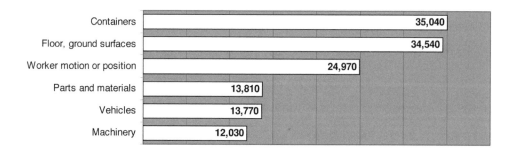

FIGURE 2.11
From Safety + Health, March 2008.

Shocking as they are, these attention-grabbing statistics do not accurately reflect the whole story. In addition to the high numbers of claims and their costs, there are indirect costs that can include the following:

- Loss of productivity of the injured worker
- Costs of hiring and training replacement workers
- Administrative costs of settling and paying the claim
- Lost sales
- Loss of customer loyalty

Office Buildings

A survey of current and emerging legal liability concerns impacting real estate managers, cosponsored by the Institute of Real Estate Management (IREM) and the National Association of Realtors (2007), found that premises liability issues, from slip-and-fall accidents to crimes on the property, were of the highest concern. The most significant current legal problems identified by the 80 IREM leaders surveyed are those relating to the day-to-day business of managing properties.

A majority of respondents (57%) ranked slips and falls as the single-leading cause of current disputes, with more than 6 in 10 (63%) ranking such accidents among the top three management issues they and their colleagues face.

Spotlight on Retail Trade
What caused your injury?
LEADING SOURCES OF INJURY OR ILLNESS 2005

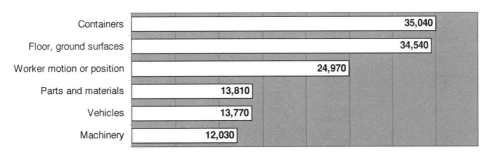

FIGURE 2.12
From Safety + Health, April 2008.

Of 847 cases and jury verdict reports analyzed, those addressing premises liability issues constituted nearly 60% of the total. Most of the cases involved slip-and-fall accidents or other injuries caused by conditions of a tenant's property or common areas.

Slips and falls account for one-third of all building insurance costs. Statistics show that the majority (60%) of falls happen on the same level, resulting from slips and trips. The remaining 40% are falls from a height.

Convenience Stores

Slips and falls are the leading cause of both frequency and severity loss issues for convenience store operations affecting both customers (general liability) and employees (workers' compensation).

The Federated Mutual Insurance Company completed a two-year study of the industry's losses that showed that slips, trips, and falls account for 39% of all convenience store general liability losses. For many convenience stores, customer slips and falls account for up to 50% of their firms' general liability losses. The study of customer slips and falls also revealed that two-thirds of these accidents occurred outside the store, while only one-third involved customers inside the store. The most likely place for a customer to slip, trip, or fall was in the store parking lot and on drive areas. Slippery surfaces due to snow, ice, water, and oil were the causes most frequently listed.

All workers' compensation claims reported to Federated by petroleum and convenience store marketers in 1997 and 1998 were also analyzed to pinpoint the impact of employee slips and falls. A slip, trip, or fall accident was shown to be twice as likely to cause an injury to a store employee than lifting. With the average cost to settle these losses running several thousand dollars each, it makes sense to check out the benefits of this new employee training program on preventing slips, trips, and falls.

Four fatalities resulted from slips and falls during the two-year loss study. Three customers died within days of falling at convenience stores: one from a blood clot and two from head injuries resulting from falls. One employee suffered a fatal fall after slipping from a ladder while hanging signs.

Coin-Operated Laundries

Slips and falls are the leading cause of guest injury in the coin-operated-laundry industry. "People come to the laundry in flip-flops, slippers, and a variety of other shoes that can contribute to a fall. People are of various size and age, which can also have an effect on the seriousness of an injury. Unattended and/or 24-hour stores do pose a greater risk than a store with normal hours or one with full-time attendants. A fall can seem much worse if no one is there to witness it or to assist the injured person with the care that is needed."

Transit Terminals

According to the U. S. Department of Transportation, Federal Transit Administration (1985), statistical analysis of more than 1,000 pedestrian falling accidents in transit stations reveals the following:

- There are about 20.7 falling accidents and eight ambulance-aided cases for each 10 million station uses.
- Falls in transit are not significantly different than other types of exposures and are fewer than the number of falls in the home.
- Younger and older age-groups have greater-than-average falling experiences.
- Alcohol involvement is a significant cause of falls, observed in 29% of all reported transit station falls and 55% of male falling incidents where an ambulance was required.
- Off-hour and weekend falling accidents are above average when compared to passenger activity and falls in the P.M. peak period are twice that for the A.M. peak.
- Escalator falls are more common but typically less severe than stair or walking surface falls.
- Approximately 90% of reported stairs falls are in the down direction.
- The most common injury location for male falls is the head and for females the legs.

It was found that although few transit falling accidents are caused by design or operating deficiencies, increased liability associated with slips, trips, and falls were most likely caused by inappropriate design or poor housekeeping.

Health Care

The health care industry is the largest employer in the United States (about 13 million workers) and ranks second among eight industries as having the highest percentage of claim costs associated with falls on the same level (Cotnam et al. 2000). In 2002, hospitals became the industry with the greatest number of total injuries (over 296,000) in the United States (Wiatrowski 2004). In addition to the large workforce and the large number of injuries generated, hospitals also have a much higher than average rate of slips, trips, and falls on the same level. In 2007, the Bureau of Labor Statistics reported that the incidence rate of lost workday injuries from slips, trips, and falls on the same level in hospitals was 35.2 per 10,000 full-time equivalents (FTEs), which was 75% greater than the average rate for all other private industries combined (20.2 per 10,000 FTEs).

Slips, trips, and falls due to liquid contamination (water, fluid, slippery, greasy, and slick spots) were the most common cause (24%) of claims for the entire study period

1996–2005. Food service, transport/emergency medical service, and housekeeping staff were at highest risk of slip, trip, and fall claims in the hospital environment, while nursing and office administrative staff generated the largest numbers of slip, trip, and fall claims.

Recent Bureau of Labor Statistics data indicate that total slips, trips, and falls accounted for 22% of all nonfatal occupational injuries and illnesses involving days away from work in 2006 (U.S. Department of Labor 2009). National Electronic Injury Surveillance System (U.S. Department of Health and Human Services 2009) data show that total slip, trip, and fall–related injuries account for the second-greatest proportion (30%) of all work-related emergency department visits requiring hospitalization. Other databases show that slips, trips, and falls account for a substantial portion of workplace injuries both in the United States and in other developed countries, as same-level slips, trips, and falls have increased in recent years, according to the Liberty Mutual Workplace Safety Index ("Making Strides in Slips and Falls Prevention from Research to Reality" 2007).

According to the 2006 Liberty Mutual Workplace Safety Index, the annual direct cost of disabling occupational injuries due to slips, trips, and falls is estimated to exceed $11 billion. The index reports that falls on the same level are the second most costly occupational injury (estimated annual cost of $6.7 billion), just behind overexertion. It also shows that bodily reaction, which comprises injuries from slipping or tripping, is the third-highest injury category, followed by falls to a lower level (4.6 bil-

Spotlight on Professional and Business Services
What caused your injury?
LEADING SOURCES OF INJURY OR ILLNESS 2005

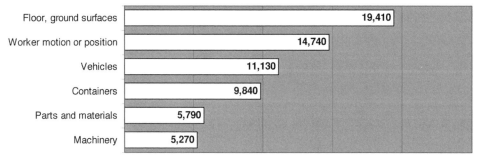

FIGURE 2.13
From Safety + Health, January 2008.

lion). Don Ostrander, director of consulting services of occupational safety and health at the National Safety Council in Itasca, Illinois, states that "the average cost from slips and falls is $22,800 per accident" and that the average workers' compensation claim is $19,000" (Zudonyi 2008).

Slip, trip, and fall injury events in hospitals have myriad causes, and the work conditions in hospitals are diverse. This research provides evidence that implementation of a broad-scale prevention program can significantly reduce slip, trip, and fall injury claims.

Nursing Homes

According to the CDC (2009), falls are often the result of underlying health problems. Nursing home residents are generally more frail than seniors living in the community and have more cognitive impairments which limit their daily living:

- In 1997, 1.5 million persons age 65 and older lived in nursing homes (Kramarow et al. 1999). If current rates continue, by 2030 this number will rise to about 3 million (Sahyoun et al. 2002).
- Each year, a typical 100-bed nursing home reports 100 to 200 falls. Many others go unreported (Rubenstein 1997).
- As many as 75% of nursing home residents fall annually (Rubenstein, Josephson, and Robbins 1994), twice the rate of seniors living in the community.
- Patients often experience multiple falls—2.6 falls per person per year on average (Rubenstein et al. 1990).
- About 35% of fall injuries occur among nonambulatory residents (Thapa et al. 1996).
- About 20% of all fall-related deaths among older adults occur among the 5% who live in nursing homes (Rubenstein 1997).
- About 1,800 fatal falls occur among residents of U.S. nursing homes each year (Rubenstein et al. 1988).
- Approximately 10% to 20% of nursing home falls cause serious injuries; 2% to 6% cause fractures (Rubenstein et al. 1988).
- Falls can result in decreased physical functioning, disability, and reduced quality of life. Loss of confidence and fear of falling can lead to further functional decline, depression, feelings of helplessness, and social isolation (Rubenstein et al. 1994).

Construction/Mining Industry

Construction industry workers are particularly vulnerable to loss of life as the result of slips, trips, and falls. About 50% of all work-related fall deaths involve construction workers. In addition, one in every three construction fatalities is due to a fall. Falls to

Spotlight on Educational and Health Services
What caused your injury?
LEADING SOURCES OF INJURY OR ILLNESS 2005

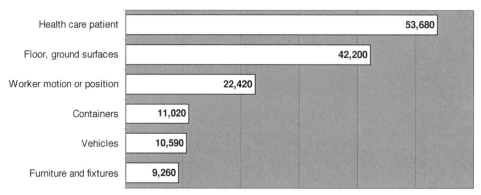

Health care patient	53,680
Floor, ground surfaces	42,200
Worker motion or position	22,420
Containers	11,020
Vehicles	10,590
Furniture and fixtures	9,260

FIGURE 2.14
From Safety + Health, February 2008.

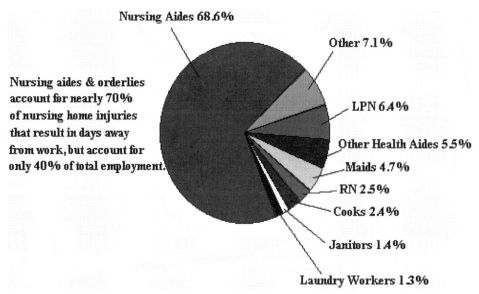

Nursing Aides 68.6%

Other 7.1%

Nursing aides & orderlies account for nearly 70% of nursing home injuries that result in days away from work, but account for only 40% of total employment.

LPN 6.4%

Other Health Aides 5.5%

Maids 4.7%

RN 2.5%

Cooks 2.4%

Janitors 1.4%

Laundry Workers 1.3%

FIGURE 2.15
From 1994 BLS Survey of Occupational Injuries & Illnesses.

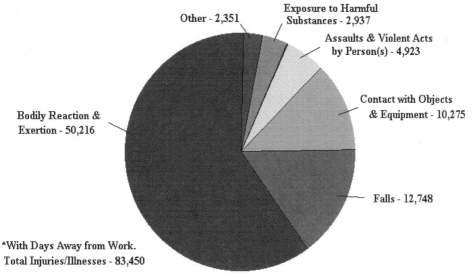

Other - 2,351

Exposure to Harmful
Substances - 2,937

Assaults & Violent Acts
by Person(s) - 4,923

Bodily Reaction &
Exertion - 50,216

Contact with Objects
& Equipment - 10,275

Falls - 12,748

*With Days Away from Work.
Total Injuries/Illnesses - 83,450

FIGURE 2.16
From 1994 BLS Survey of Occupational Injuries & Illnesses.

Spotlight on Construction
What caused your injury?
LEADING SOURCES OF INJURY OR ILLNESS 2005

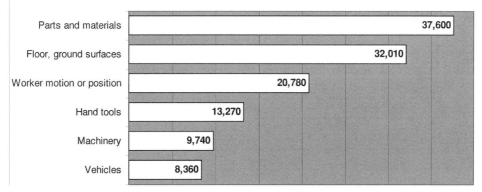

Parts and materials	37,600
Floor, ground surfaces	32,010
Worker motion or position	20,780
Hand tools	13,270
Machinery	9,740
Vehicles	8,360

FIGURE 2.17
From 1994 BLS Survey of Occupational Injuries & Illnesses.

a lower level are the most likely to result in a fatality, accounting for nearly 90% of work-related deaths. These falls are further broken down into falls from ladders, roofs, and scaffolds and falls to the same level, with the first three categories accounting for most deaths. According to the NSC, from 1993 to 1997, 23% of construction injuries/illnesses involving days away from work were attributed to slips, trips, and falls. Slips, trips, and falls also account for a significant proportion of nonfatal occupational injuries, which in 1998 accounted for 20% of the injuries/illnesses involving days away from work..

Workers in certain segments of the mining industry are also at risk of slips, trips, and falls. According to the Mine Safety and Health Administration, between 1996 and 1999, 28 miners died from slips or falls. Some 86% of these fatalities occurred in operations other than coal mining (e.g., metal, nonmetal, stone, sand, and gravel). The greatest proportion (40%) of fall fatalities has occurred in stone mining—an industry that accounts for about 24% of the average total employment in the mining industry.

From 1996 to 1999, more contractors died from falls in the stone mining sector than contractors in all other sectors combined.

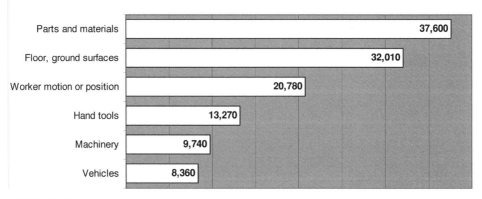

Spotlight on Construction
What caused your injury?
LEADING SOURCES OF INJURY OR ILLNESS 2005

Source	Value
Parts and materials	37,600
Floor, ground surfaces	32,010
Worker motion or position	20,780
Hand tools	13,270
Machinery	9,740
Vehicles	8,360

FIGURE 2.18
From Safety + Health, July 2008.

Contract workers account for more than half the fatalities (53%) recorded. Slips, trips, and falls also accounted for 22% of nonfatal lost-time injuries. However, 92% of these injuries occurred in nearly equal proportions across mining sectors. The greatest proportion (43%) occurred in coal mining. In every sector, the number of miners experiencing nonfatal lost-time fall-related injuries is far greater than the number of contractors who experience such injuries.

Trucking Industry

According to the American Trucking Association, slips and falls are the leading cause of compensable injury in the trucking industry. Slip-and-fall accidents are a significant factor in driver injuries, causing approximately 22% of injuries to drivers. Slips and falls are the most expensive type of driver accident other than being involved in a motor vehicle collision. An analysis of a driver's job activities reveals several distinct functions, each of which presents unique exposures for slip-and-fall accidents. A driving employee is normally involved in the following:

1. Operating the vehicle (including entering and exiting from the vehicle)
2. Handling freight (including loading and unloading merchandise from the vehicle)
3. Routine walking and climbing around the truck terminal or customer's place of business

Other Industries

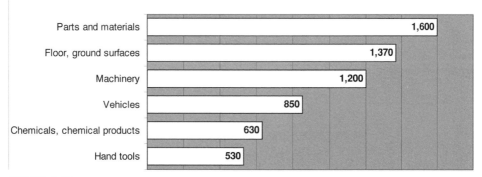

FIGURE 2.19
From Safety + Health, November 2008.

Spotlight on Utilities
What caused your injury?
LEADING SOURCES OF INJURY OR ILLNESS 2005

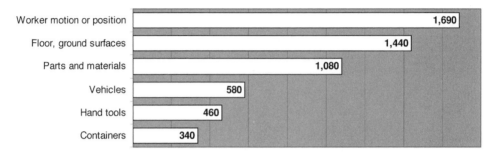

Worker motion or position	1,690
Floor, ground surfaces	1,440
Parts and materials	1,080
Vehicles	580
Hand tools	460
Containers	340

FIGURE 2.20
From Safety + Health, June 2008.

Spotlight on Agriculture, forestry, fishing and hunting
What caused your injury?
LEADING SOURCES OF INJURY OR ILLNESS 2005

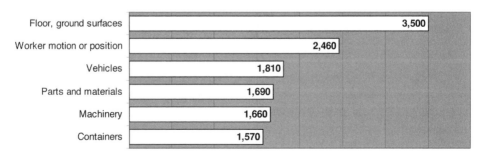

Floor, ground surfaces	3,500
Worker motion or position	2,460
Vehicles	1,810
Parts and materials	1,690
Machinery	1,660
Containers	1,570

FIGURE 2.21
From Safety + Health, December 2008.

BATHROOM/BATHTUB SLIPS AND FALLS

A recent study conducted by the Zurich Insurance Company found that 41% of their general liability and workers' compensation claims were the result of slips and falls. Of these, 32% are due to moisture on bathroom floors. Many include broken bones, back and head injuries, and even fatalities. Their data revealed that the bathroom is a common area for slip-and-fall accidents because floors are often wet, soapy, and slippery. They found that slips and falls in bathtubs and showers are the most common types of accidents and that the floor surface is the single most important factor that contributes to a slip-and-fall accident.

According to a study published in the journal *Pediatrics* (August 2009), 120 children require emergency room treatment each day as a result of a bathtub or shower slip and fall. Most injuries are to children under the age of four. The study was based on a nationally representative survey of bathtub and shower–related injuries to children 18 years and younger in the United States.

An average of more than 43,000 children are treated in hospital emergency departments annually for bathtub and shower–related injuries. Children age four and younger accounted for more than half of all bathtub and shower–related injuries. About 60% of the injuries were lacerations (cuts and tears to the skin), and more than 20% were sprains and other soft-tissue injuries. The most commonly affected body region was the head, accounting for nearly 50% of the bath injuries, followed by the head and neck, which accounted for 15% of the injuries.

Overall, wounds from falls make up about 80% of bath-related injuries, topping those caused by scalding water or submersion in the tub. In most cases, parents were watching their kids at the time.

ESCALATOR FALLS

The CPSC estimates that 90 billion riders will use an escalator, most without incident. However, the CPSC estimates there were approximately 11,000 escalator-related injuries in 2007. The majority of these injuries are from falls, but 10% occur when hands, feet, or shoes are trapped in escalators. The CPSC (2001) estimated that there are 6,000 hospital emergency room–treated injuries associated with escalators each year.

Information on passenger injuries and deaths is reported through the CPSC's NEISS. During the nearly 10-year study (1992–2003), the CPSC reported 24 non–work-related deaths of escalator passengers in 12 states and the District of Columbia—about two per year. The states were Alabama (one death), California (two), District of Columbia (three), Florida (one), Illinois (three), Maryland (one), Minnesota (three), Nevada (one), New York (three), Ohio (one), Virginia (one),Washington (two), and Wisconsin (two). The eight "caught in/between" deaths usually resulted after clothing became trapped at the bottom or top of an escalator or between a stair and escalator sidewall; seven of the 16 fall deaths were from head injury. Four of the fall deaths occurred from falling off the escalator while riding the escalator side rails.

In 1994, the CPSC estimated that there were 7,300 escalator and 9,800 elevator injuries requiring hospitalization (Cooper 1997; CPSC 1998). The data were based on a nationwide survey of 90 hospitals. Based on the number of elevators and escalators in the United States, the CPSC estimated that there were 0.221 accidents per escalator and 0.015 accidents per elevator annually.

The CPSC estimated that 75% of the escalator injuries resulted from falls, 20% from entrapment at the bottom or top of an escalator or between a moving stair and escalator sidewall, and 5% "other." The "caught-in" incidents generally resulted in more serious injuries than did falls. Of particular concern is the fact that half of the approximately 1,000 sidewall-entrapment injuries involved children under age five (Armstrong 1996b). The children's injuries were caused mostly when a child's hands or footwear (including dangling shoelaces) became caught in an escalator comb plate at the top or bottom of an escalator or in the space between moving stairs and an escalator sidewall.

The CPSC is aware of 77 entrapment incidents since January 2006, with about half resulting in injury. All but two of the incidents involved popular soft-sided flexible clogs and slides.

CPSC REPORTS: ESCALATOR PASSENGER INJURIES AND DEATHS

- A 37-year-old male died from asphyxiation when his clothing became entrapped in the downward-moving steps and stationary bottom comb plate of an escalator at a subway station. He was found, on his back, with the coat wrapped tightly around his chest because part of the coat was dragged into the comb plate. There were no witnesses as to how the coat became entangled (March 11, 1997, Washington, D.C.).

- A female, age 85, lost her balance and fell onto the escalator at a store. Cause of death was a blunt impact to head, trunk, and extremities sustained in the fall (September 11, 2000, Richmond Heights, Ohio).

- A 12-year-old male was riding an escalator down (egress) from a baseball game when his right shoe got stuck between the stationary left side of the escalator. The victim sustained injury to his right big toe. The extent of the injury was not determined (July 6, 2002, Anaheim, Calif.).

- A 5-year-old female was on the bottom step of a down escalator when her shoe got caught in the comb plate. She reached down to get her shoe when her hand also got caught in the comb plate. Her three middle fingers and part of her hand were amputated (February 19, 2003, St. Petersburg, Fla.).

- An escalator incident at a theater caused 71 children to suffer minor injuries. The escalator was heading up when it abruptly stopped and shifted slightly backward, causing the kids to fall down (January 13, 2005, New York City).
- A 16-month-old girl was injured when her hand slipped between the moving stairs and the escalator wall, and she became lodged. She may have been in a stroller when the incident happened. She was flown to the hospital for surgery (March 13, 2006, Glendale, Ariz.).

Each year, countless numbers of passengers are injured as a result of falling down escalators, most of which go unreported. Malfunctioning escalators are also a cause of death or injury when an escalator suddenly sped up or reversed direction (Armstrong 1996a).

According to an Indiana University study (O'Neil and VanSwearingen 2008), the rate of injuries to older adults riding escalators more than doubled from 1991 to 2005. Reviewing records from CPSC, the study found that nearly 40,000 adults age 65 and older were injured on escalators between 1991 and 2005. In fact, the rate rose more than twofold during that period: from 4.9 injuries per every 100,000 older Americans in 1991 to 11 injuries per 100,000 people by 2005. The report stated that "the trend of increasing escalator accidents is likely related to shifts in lifestyle. Older adults are now

Spotlight on Information
What caused your injury?
LEADING SOURCES OF INJURY OR ILLNESS 2005

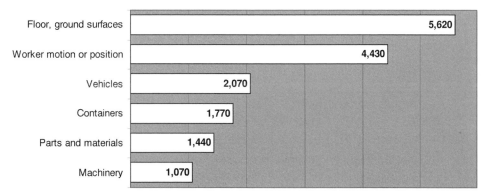

FIGURE 2.22
From Safety + Health, October 2008.

more active at an older age than probably ever before and the mean age of the accident victims in the study was 80 years old."

Although the injury numbers were growing, most of the injuries were not serious. Only 8% of the injured were admitted to a hospital after evaluation in an emergency department. The most common injuries were to the lower extremities (about 26%) and the head (25%). The leading type of injury was soft-tissue injuries (54%), such as sprains, followed by cuts (about 22%) and fractures (almost 16%). Women accounted for more than 73% of escalator injuries. The most frequent cause of escalator injury was a slip, trip, or fall, accounting for 85% of all injuries.

AN INOPERABLE (STATIONARY) ESCALATOR IS NOT A STAIRWAY!
A stationary escalator is not a stairway and should not be used as one. Most escalator manufacturers warn their customers of this, and many simply ignore the warning. An inoperative escalator should be barricaded from use until it becomes operable.

The only recognized escalator standard is published by the American Society of Mechanical Engineers (ANSI/ASME A17.1 "Safety Code for Elevators and Escalators"). This standard discusses the importance of applying lines of demarcation along the sides and backs of each step to draw the user's attention to these locations so as to prevent injury, but it does not address the use of a stationary escalator.

In 1892, George A. Wheeler of New York City patented ideas for the first practical moving staircase, though it was never built (U.S. No. 479864). Some of its features were incorporated in the prototype built by the Otis Elevator Company in 1899. Charles D. Seeberger coined the brand name Escalator (from *scala*, Latin for steps, with *elevator*). Seeberger with Otis installed the first step-type escalator made for public use at the Paris Exhibition of 1900, where it won first prize. Seeberger eventually sold his patent rights to Otis in 1910. The earliest type of escalator, patented in 1891 by Jesse W. Reno, was introduced as a new novelty ride at Coney Island, moving passengers on a conveyor belt at an angle of 25 degrees.

FALLS PREVENTION INTERVENTIONS IN THE MEDICARE POPULATION

Falls are a significant health problem for elderly adults. One in three people age 65 and older and 50% of those 80 and older fall each year. Falls can have devastating outcomes, including decreased mobility, function, and independence and, in some cases, death. Health care for fall-related injuries is expensive. One estimate suggests that direct medical costs for fall-related injuries was $20.2 billion in 1994 and will rise

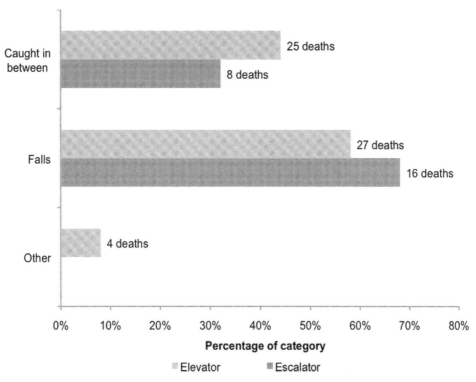

FIGURE 2.23
From Consumer Product Safety Commission.

to $32.4 billion by 2020. Another suggests that these costs will reach $240 billion by 2040.

With nearly 40 million baby boomers about to reach retirement age, the number of age-related slips and falls will more than double by 2020. The most common injury as the result of a fall is a broken hip.

The United States is about to experience a great demographic shock. Between 2010 and 2030, the over-65 population will increase more than 70%, while, under current law, the population paying payroll taxes will increase less than 4% (Congressional Budget Office 1997). The implied shortage of workers and the rapid increase in the number of retirees has profound implications for the private sector and the amount of economic growth that can be expected in the future. It also has important implications for public policies affecting the economic and physical well-being of the elderly population.

REPORT BREAKS DOWN THE CAUSES OF DEATHS IN IOWA
A University of Iowa report finds an average of four Iowans are killed every day, or more than 1,500 people a year, by injuries. John Lundell, deputy director of

the University of Iowa Injury Prevention Research Center, says that the first-of-its-kind report is broken down by age-group, gender, and injury and covers the years 2002 through 2006 (Matt Kelley, Radio Iowa News, March 27, 2009).

While many young people can trip and fall with no consequences other than maybe a skinned knee, Lundell says that such is not the case for the older generation. Falls are a significant cause of death in the elderly. Lundell says, "Many of us have experienced a fall of a loved one that might result in a hip fracture or a head injury and oftentimes, unfortunately, it's the cause that finally puts them to their final rest." The report also looks at Iowans' injuries geographically.

Lundell says that smaller counties, those with populations of less than 10,000, had the highest injury-death rate, while counties with populations of more than 50,000 had the lowest injury-death rate. While injuries kill an average of four Iowans a day, Lundell says that for every person who dies, another 250 make emergency room visits.

The United States Is Not Alone!

Like most of western Europe, the United States is becoming a nation of the elderly. America's elderly population is now growing at a moderate pace. But not too far into the future, the growth will become rapid—so rapid, in fact, that by the middle of the century, it might be completely inaccurate to think of ourselves as a nation of the young: there could be more persons who are elderly (65 or over) than young (14 or younger). With this comes the issue of who will pay for the care of an aging society. By 2050, the ratio of inactive (unemployed) elderly people to that of active (employed) individuals may be one to one (50%). For every worker paying into the Medicare/Medicaid system, one will be taking money from the system.

In Great Britian, over a third of all major injuries reported each year are caused as a result of a slip or trip—the single most common cause of work injury—costing employers over £512 million a year in lost production and other costs. Slips and trips also account for over half of all reported injuries to the British general public.

The elderly population has grown substantially. During the twentieth century, the number of persons in the United States under age 65 tripled. At the same time, the number aged 65 or over has jumped by a factor of 11. Consequently, the elderly, who comprised only 1 in every 25 Americans (3.1 million) in 1900, made up 1 in 8 (33.2 million) in 1994. Declining fertility and mortality rates also have led to a sharp rise in the median age of the nation's population—from 20 years old in 1860 to 34 in 1994 (Economics and Statistics Administration, 1995).

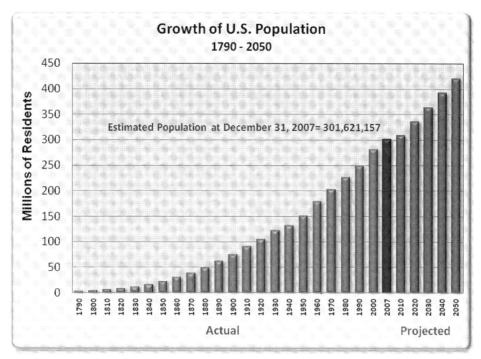

FIGURE 2.24

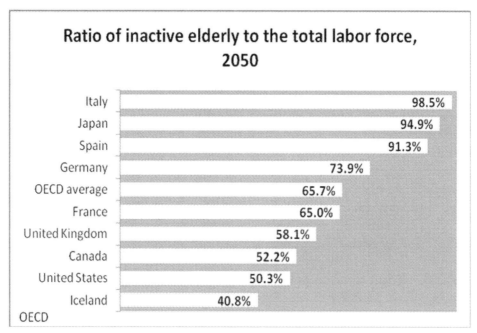

FIGURE 2.25
From OECD Factblog.

During the period 1990–2010, the elderly growth rate will be lower than during any 20-year period since 1910, a result of the low fertility of the 1930s. After this slow-growth period, an elderly population explosion between 2010 and 2030 is inevitable as the baby-boom generation reaches age 65. During that period, the number of elderly will grow by an average of 2.8% annually. By comparison, annual growth will average 1.3% during the preceding 20 years and 0.7% during the following 20 years. About one in five U.S. citizens will be elderly by 2030. The elderly population numbered 30 million in 1988, will not reach 40 million until 2011, and then will reach 50 million in only eight years (2019). According to the Census Bureau's "middle series" projections, the elderly population will more than double between now and the year 2050 to 80 million.

Ten years from now, the elderly population of the United States will double in number.

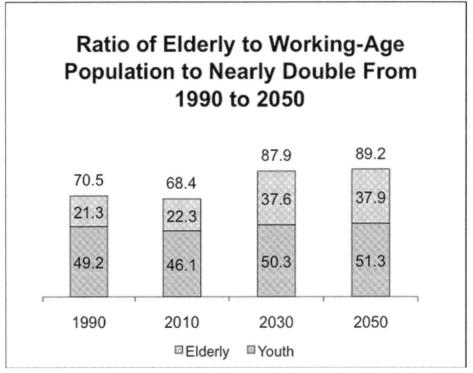

FIGURE 2.26
From United States Census Bureau.

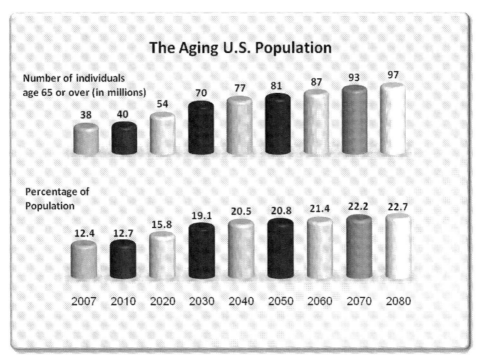

FIGURE 2.27

The "oldest old"—those aged 85 and over—are the most rapidly growing elderly age-group. Between 1960 and 1994, their numbers rose 274%. In contrast, the elderly population in general rose 100%, and the entire U.S. population grew only 45%. The oldest old numbered 3 million in 1994, making them 10% of the elderly and just over 1% of the total population. Thanks to the baby-boom generation, it is expected that the oldest old will number 19 million in 2050. That would make them 24% of the elderly Americans and 5% of all Americans. Under the "highest" projection series, the oldest old could number as many as 31 million in 2050. Since the oldest old often have severe chronic health problems that demand special attention, the rapid growth of this population group has many implications for individuals, families, and governments:

- The elderly population will grow faster than the total population.
- Growth in the 65-and-older population will outstrip population growth by a factor of 3.5.
- Warmer states, such as Nevada, Arizona, California, Florida, North Carolina, and Texas, will attract the elderly population and will grow substantially.
- Over half the states in the United States will double their elderly populations by 2030.

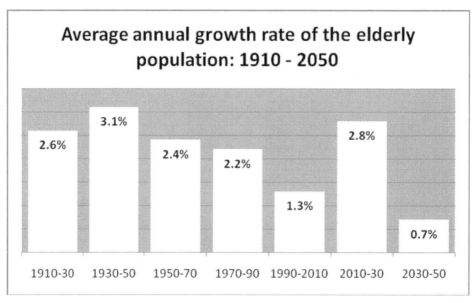

FIGURE 2.28

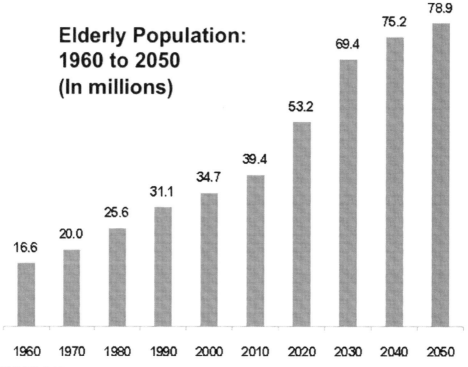

FIGURE 2.29
From United States Census Bureau.

In 2004, Medicare expenses exceeded revenues by nearly $3 billion—a first in the program's history—and Medicare's expenses are projected to grow at nearly 7% per annum for the foreseeable future, according to a recent actuarial report released by the federal government (Old-Age, Survivors, and Disability Insurance Program 2005). Seven percent is substantially greater than the growth of the gross domestic product (i.e., tax revenue growth), making Medicare's annual expense growth inherently unsustainable.

While the aging of the population may give the elderly and even certain states more political clout, the math doesn't work, and increased Medicare expenses will simply have to be curtailed if taxes are not raised (HealthPointCapital.com).

We're Living Longer

Back when the United States was founded, life expectancy at birth stood at only about 35 years. It reached 47 years in 1900, jumped to 68 years in 1950, and steadily rose to 76 years in 1991. From 1900 to 1960, the elderly increased 10-fold, while the population under age 65 was only 2.2 times larger. Between 1960 and 1990, the elderly grew by 88%, compared to 34% for persons under age 65. In 1991, life expectancy was higher for women (79 years) than for men (72 years). During the 1980s, life expectancy improved for all race/sex groups. The biggest improvement (a rise of over one year) belonged to white men.

> **MORE OLDER WOMEN**
>
> One of the most striking characteristics of the older population is the change in the ratio of men to women as people age. In 2002, 26.6 million men and 33.0 million women in the civilian noninstitutionalized population were aged 55 and over, yielding a sex ratio (men per 100 women) of 81. The sex ratio drops steadily with age. In the 55-to-64 age-group, the sex ratio was 92, and in the age-group 85 years and over, the sex ratio was 46.

WHO IS MORE LIKELY TO FALL—MEN OR WOMEN?

A sample of 761 subjects 70 years and over was drawn from general-practice records of a rural township. Each subject was assessed and followed for one year to determine the incidence of and factors related to falls. The fall rate (number of falls per 100 person-years) increased from 47 for those aged 70 to 74 years to 121 for those 80 years and over. There was no sex difference in fall rate, but men were more likely than women to fall outside and at greater levels of activity. Twenty percent of falls were associated with trips and slips, but we found no evidence that inspection of homes and installation of

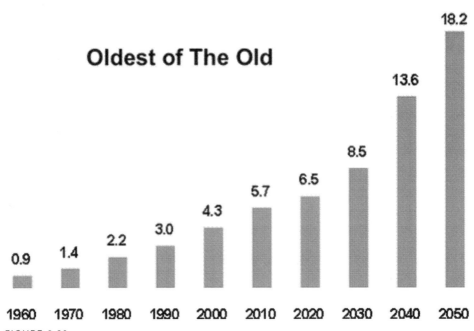

FIGURE 2.30
From United States Census Bureau.

safety features would have decreased the fall rate. Ten percent of falls resulted in sig-nificant injury. Men who fell had an increased subsequent risk of death compared with those who did not fall (relative risk 3.2, 95% confidence interval 1.7–6.0). Subsequent mortality was increased among women who fell, but not to significant levels (relative risk 1.6, 95% confidence interval 0.9–2.7).

Injuries as a Result of Falls

The most common fall-related injuries are osteoporotic fractures. These are frac-tures of the hip, pelvis, femur, vertebrae, humerus, hand, forearm, leg, and ankle. In 1986 in the United States, the overall direct medical costs of osteoporosis were $15.5 billion (Norris 1992). More recent estimates put the costs at $10 billion (Barrett-Con-ner 1995). By 2000, the economic impact of osteoporotic fractures was expected to exceed $45.2 billion (Chrischilles et al. 1994).

Canes and Walkers Contribute to Elderly Falls

Researchers at the Centers for Disease Control and Prevention examined emergency room medical records from 2001 to 2006 and found that an average of 129 Americans age 65 and older were treated each day in emergency rooms for fall-related injuries involving walkers and canes. Fractures were the most common injury, and one out of every three injuries required hospitalization. Older adults were seven times more likely to be injured in a fall with a walker than with a cane. Older women sustained 78 percent of walker injuries and 66 percent of all cane-related injuries. The study concluded that older adults need to use walkers and canes more safely.

Typical Elderly Fall Injuries: Clinical Scenario (Source: Howard W. Harris, M.D.)

- Mr. Slip is a 78-year-old, right-hand-dominant male who lives alone. He presented to the emergency room after falling in his garage while taking out the trash. He has had two other falls over the past year and a half. He is scared of falling again. He has a history of osteoarthritis and anxiety/depression. He is on naproxen and diazepam. He has family who lives close by and helps watch after him.
- Mrs. Fall is an 81-year-old female who was found down in her bathroom after not showing up for breakfast at her retirement community this morning. She was transported to the emergency room for altered mental status, deformity, and pain in the left lower extremity. She has a history of mild dementia and atrial fibrillation, for which she is on digoxin and Coumadin (a blood-thinning agent).

Medical Significance of Slips and Falls

The medical significance of slips and falls can be characterized as follows:

- Most common cause of nonfatal injuries annually (20% to 30% of all falls)
- Most common cause of hospital admissions for trauma
- More than 388,000 hospital admissions annually
- Consequences of a fall
- Fractures result in 6% of falls (400,000–500,000)
- Hip, vertebrae, forearm, leg, ankle, pelvis, humerus, and hand
- Head injury—subdural hematoma, traumatic brain injury
- Soft-tissue injury
- Fear of falling can result in decreased activity, isolation, and further functional decline
- Nursing home placement and loss of independence (four to five times greater risk)
- Increased dependence on family and society

Hip Fractures

Hip fracture data are as follows:

- 1% of falls result in hip fracture.
- 90% of hip fractures are from a fall (more than 340,000 hospital admissions annually).

- Risk exponentially increases with age.
- More than $10 billion in medical costs are incurred annually.
- 60% have long-term restricted mobility.
- 25% remain functionally more dependent.
- 25% die within six months.
- In 2040, it is estimated that over 500,000 hip fractures will be caused by falls.

Risk Factors and Etiology Dysmobility

Data on risk factors and etiology dysmobility are as follows:

- Dysmobility and falling are closely related.
- 15% over age 65 have trouble walking.
- 25% of men and 33% of women over age 85 have difficulty walking.
- Two-thirds of people in hospitals or nursing homes are unable to ambulate without assistance.
- 50% to 60% of all falls occur at home.
- The majority of falls occur on the same level.
- The majority of falls occur while standing.
- 40% to 50% of fall-related work injuries are attributable to slipperiness or slipping.

The survey also found that more than a quarter (26%) of U.S. adults said that they are not worried that an injury may occur in their own home.

Angela Mickalide, director of education and outreach for the Home Safety Council (2007), said, "Each year in our nation home-related injuries result in nearly 20,000 deaths and 21 million medical visits, many of which are almost entirely avoidable with proper education and a few simple home modifications."

Falls Prevention Key Findings

When asked to identify which type of injury they are most worried might happen in their own home, only about one-fifth (19%) of survey respondents were concerned about falls—the leading cause of home injury death:

- Slips and falls are a serious cause of injury and death in both the working and non-occupational environments. Total (occupational and nonoccupational) U.S. fatalities due to falls range between 12,000 and 15,000 per year, second only to motor vehicle crashes as the accidental cause of death. In 1995, there were approximately 14,000 fatalities in the United States due to falls, compared to 43,000 fatalities in motor vehicle accidents. In the nonoccupational environment, victims are typically the very young and the very old. Programs to reduce falls in infants and toddlers (e.g., safety gates and warning labels on walkers) have been successful in reducing fatalities during the 1990s.

- There are approximately 600 occupational fatalities due to falls each year, a 50% reduction since the 1970s.
- Total injuries due to falls are estimated at 13 million to 14 million per year in the United States. Falls are the leading cause of accidental injury, resulting in 20.8% of all emergency room visits in 1995. (Motor vehicle accidents accounted for 11.9% of emergency room visits.)
- Disabling (temporary and permanent) occupational injuries due to falls are approximately 250,000 to 300,000 per year.
- Falls occur in virtually all manufacturing and service sectors. Fatal falls, however, are in construction, mining, and certain maintenance activities.
- According to workers' compensation statistics from ITT-Hartford Insurance Company, falls account for 16% of all claims and 26% of all costs. This compares to 33% of costs associated with sprains and strains.
- According to the American Trucking Association, slips and falls are the leading cause of compensable injury in the trucking industry.
- Falls are generally classified into one of two categories (ANSI "Accident Type"): (1) falls from elevation (approximately 40% of compensable fall cases and approximately 10% of occupational fatalities) and (2) falls on the same level (approximately 60% of compensable fall cases) (Keyserling 2000).

REFERENCES

Armstrong, David. 1996a. "Escalator Accident Injures 22." *Boston Globe*, February 22.

———. 1996b. "US Urges Upgrade in Elevator Safety." *Boston Globe*, July 21.

Barrett-Connor, E. "The Economic and Human Costs of Osteoporotic Fracture." *American Journal of Medicine* 98, suppl. 2A (1995): 2A–3S to 2A–8S.

Centers for Disease Control and Prevention, National Institute for Occupational Safety and Health. DHHS (NIOSH) Publication No. 2009-139, August 2009.

Chrischilles, E., T. Shireman, and R. Wallace. "Costs and Health Effects of Osteoporotic Fractures." *Bone* 15, no. 4 (1994): 377–86.

Coalition Against Insurance Fraud. "How Big Is $80 Billion?" http://www.insurancefraud.org, 2008.

Congressional Budget Office. *Long-Term Budgetary Pressures and Policy Options*. Washington, DC: U.S. Government Printing Office, March 1997.

Consumer Product Safety Commission. 1998. "Escalator Safety." CPSC Document #5111. http://www.cpsc.gov/cpscpub/pubs/5111.html (1998).

———. 2001. "Consumer Product Safety Alert: Escalator Safety." http://www.cpsc.gov/cpscpub/pubs/5111.pdf (2001).

Cooper, David. 1997. "Escalator Side-of-Step Entrapment." Paper presented at International Association of Electrical Engineers Elevcon '96, Barcelona, Spain. http://www.elevator-expert.com/escalato.htm (1997).

Cotnam, J. P., W. R. Chang, and T. K. Courtney. "A Retrospective Study of Occupational Slips, Trips, and Falls across Industries." In *Proceedings of the 44th Annual International Ergonomics Association/Human Factors Ergonomics Society Congress*, San Diego, CA, vol. 4 (2000): 473–76.

Economics and Statistics Administration, U.S. Department of Commerce. *Sixty-Five Plus in the United States.* Washington, DC: *U.S. Census Bureau, Population Division, May 1995.*

Employers Insurance Company of Wausau. "Technical Reference—Hospitality, General Liability" (SIC Code 7011) (2006a).

———. "Technical Reference—Hospitality, Workers Compensation" (SIC Code 7011) (2006b).

Finkel, Mort, and James Bitzer. "Preventing Falls at Malls." *Shopping Center Business*, May 1998, 354.

Home Safety Council. "A Safe Home Is in Your Hands." Press release, 2007.

Institute of Real Estate Management and National Association of Realtors. "Survey I.D.'s Top Legal Issues." *Multi-Housing News*, August 21, 2007.

Keyserling, W. Monroe. IOE 539 Notes: Working Surfaces/Slips and Falls, 2000.

Kohr, Robert L. *Accident Prevention for Hotels, Motels, and Restaurants.* New York: Van Nostrand Reinhold, 1991.

Kramarow, E., H. Lentzner, R. Rooks, J. Weeks, and S. Saydah. *Health and Aging Chartbook. Health, United States, 1999.* Hyattsville, MD: National Center for Health Statistics, 1999.

Lord, Stephen R., Catherine Sherrington, and Hylton B. Menz. "Falls in Older People: Risk Factors and Strategies for Prevention."

"Making Strides in Slips and Falls Prevention from Research to Reality." *Liberty Mutual Research Institute for Safety* 10, no. 3 (autumn 2007).

National Institute of Occupational Safety and Health. *Child Labor Research Needs.* NIOSH Publication No. 97-143, Washington, DC: National Institute of Occupational Safety and Health, 1997.

National Safety Council. *Injury Facts.* Itasca, IL: NSC Press, 2006.

———. *Injury Facts.* Itasca, IL: NSC Press, 2008.

Norris, R. J. "Medical Costs of Osteoporosis." *Bone* 13 (1992): S11–S16.

Old-Age, Survivors, and Disability Insurance Program. *The 2005 Annual Report of the Board of Trustees of the Federal Old-Age and Survivors Insurance and Federal Disability Insurance Trust Funds.* Washington, DC: U.S. Government Printing Office, March 23, 2005.

O'Neil, Joseph, and Jessie VanSwearingen. *Accident Analysis and Prevention,* March 2008.

Rubenstein, L. Z. "Preventing Falls in the Nursing Home." *Journal of the American Medical Association* 278, no. 7 (1997): 595–96.

Rubenstein, L. Z., K. R. Josephson, and A. S. Robbins. "Falls in the Nursing Home." *Annals of Internal Medicine* 121 (1994): 442–51.

Rubenstein, L. Z., A. S. Robbins, K. R. Josephson, B. L. Schulman, and D. Osterweil. "The Value of Assessing Falls in an Elderly Population: A Randomized Clinical Trial." *Annals of Internal Medicine* 113, no. 4 (1990): 308–16.

Rubenstein, L. Z., A. S. Robbins, B. L. Schulman, J. Rosado, D. Osterweil, and K. R. Josephson. "Falls and Instability in the Elderly." *Journal of the American Geriatrics Society* 36 (1988): 266–78.

Sahyoun, N. R., L. A. Pratt, H. Lentzner, A. Dey, and K. N. Robinson. *The Changing Profile of Nursing Home Residents: 1985–1997. Aging Trends. No. 4.* Hyattsville, MD: National Center for Health Statistics, 2001.

Thapa, P. B., K. G. Brockman, P. Gideon, R. L. Fought, and W. A. Ray. "Injurious Falls in Nonambulatory Nursing Home Residents: A Comparative Study of Circumstances, Incidence and Risk Factors." *Journal of the American Geriatrics Society* 44 (1996): 273–78.

"Two-Thirds of U.S. Workers Don't Care about Their Work." *Injury Prevention and Cost Control Alert,* December 14, 2007, 1.

U.S. Bureau of Labor Statistics. *2006 Nonfatal Occupational Injuries and Illnesses Requiring Days Away from Work.* Washington, DC: U.S. Bureau of Labor Statistics, 2006.

U.S. Department of Health and Human Services, Centers for Disease Control and Prevention, National Institute for Occupational Safety and Health. DHHS (NIOSH) Publication No. 2009-139, August 2009.

U.S. Department of Labor, Bureau of Labor Statistics. *Census of Fatal Occupational Injuries (CFOI)—Current and Revised Data 1992–2008.* Washington, DC: U.S. Department of Labor, 2009.

U.S. Department of Transportation Federal Transit Administration. "UMTA-MA-06-0098-84-2, DOT-TSC-UMTA-84-36." Port Authority of New York and New Jersey, February 1985. Final Report.

Wiatrowski, W. J. "Comparing Old and New Statistics on Workplace Injuries and Illnesses. *Monthly Labor Review,* December 2004.

Winn, Gary L., Austin L. Winn, and Elyce A. Biddle. "Observations on a Teen's First Job." *Professional Safety: Journal of the American Society of Engineers*, June 2007.

Zudonyi, Corinne. "Preventing Slips, Trips and Falls: Improving Floor Safety Protocols Can Recoup Big Dollars for Facility Managers." *Housekeeping Solutions*, February 2008, 13.

II

THE REASON

3

Floors, Mats, and Cleaners

FLOOR COVERINGS

Since floors are the leading cause of slips, trips, and falls, it is important to understand the various types of hard-surface flooring materials and their impact on safety.

It was in the floor-coverings industry that my interest in floor safety began. In 1986, I began my career as a salesman for a regional floor-covering distributor in Texas. I sold resilient floor coverings to flooring retailers who in turn sold them to you, the consumer. One of the most common questions I was asked by my customers was, "How can we make our floors less slippery?" Like any good salesperson, I went to my management for an answer, only to be told that "any questions regarding slips and falls should to be directed to our attorneys." So much for customer service. The fact was, then as now, that the floor covering industry simply didn't care about the problem of slips and falls—as long as they were not being sued, why get involved?

My firsthand knowledge regarding the real-world problem of slips and falls as faced by end users led me to start my first company. I understood that I was but a cog in a multi-billion-dollar wheel that had no desire to address the third-rail issue of slips and falls. Hard-surface floors are not manufactured to be slippery, but they become so when wet. The advice from the floor-covering industry was simple: keep your floors dry, and you will not have people slipping. The problem is that such advice is often unrealistic and impractical.

FLOORING TYPES

The three major types of hard-surface floors are resilient floors, including vinyl (sheet or tile) and laminate; ceramic tile, including stone; and natural flooring materials, such as wood. We begin with resilient flooring.

Vinyl

Although one of the most versatile and cost-effective flooring types, today's vinyl and laminate floors offer superior slip resistance. But not all resilient floors are the same. The type of floor tile you walk on when shopping at your local grocery store is not the same material you would use in your kitchen at home. The biggest difference is in the wear layer. Commercial vinyl composition tile (VCT) like that used in the retail industry does not have a protective wear layer and requires the application of a floor finish or polish, often referred to as "wax." Without the application of a floor finish,

FIGURE 3.1
Vinyl Composite Tile.

VCT will wear quickly and become difficult to clean and unsightly. The resilient-floor-covering industry claims that it need not publish slip-resistance information on its products because the consumer is not actually contacting the surface of the tile but rather the protective finish. This sounds reasonable until you look at the fact that with the exception of VCT, most resilient floors are manufactured with a protective wear layer, often referred to as "no-wax" floors.

> It may surprise you to know that there are no slip-resistance standards for most types of hard-surface floors.

Because of the absence of hard-surface slip-resistance safety standards, rarely are manufacturers of resilient floor coverings sued for negligence. But their customers are.

You can't claim that products are unsafe when there is no accepted industry standard to compare the products to. And that's just the way manufacturers want to keep it. It's an unfair system where the manufacturers consciously choose not to publish the slip-resistance data for their products because they are afraid of the liability that would be associated with such data. Furthermore, most manufacturers of resilient floors have consciously chosen to not discuss specific cleaning procedures for their floors and provide only general advice, such as recommending the use of a "neutral cleaner." Such failure to inform the consumer has directly contributed to the growing rate of slip-and-fall accidents and has left consumers to fend for themselves. Manufacturers know that the average person does not know what a neutral floor cleaner is. After all, shampoo is a neutral cleaner, right?

The fact is that most resilient floors are manufactured with a mid to high level of slip resistance and, with proper maintenance, can provide a relatively safe walking surface. However, with improper maintenance, even the safest floors can become dangerous.

CURRENT FLOOR-COVERING DATA
Although slips and falls are a significant problem often associated with hard-surface floors, carpets cover most of the rooms in the average American home:

- More than 6 in 10 U.S. home owners indicate that their bedrooms, living rooms, offices/dens, and family rooms are carpeted.

- Six in 10 also cover their hallways with carpet (59%), and more than 3 in 10 have carpeted basements (37%).

There is no single dominant flooring surface for dining rooms and kitchens:

- Thirty-nine percent of dining rooms are carpeted, 29% have wood floors, and 11% have tile floors; 29% of kitchens have tile floors, 25% have linoleum floors, and 14% have wood floors.
- Those with other hard surfaces have their own unique challenges—keeping grout clean, keeping the floors clean, cracks/chips/scratches, keeping their floors from looking new, too much dirt accumulating on the floors, and so on. Four in 10 do not have any challenges with maintaining their hard-surface flooring.

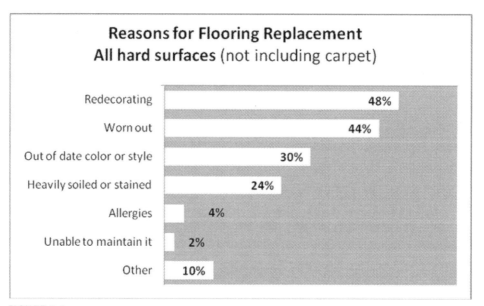

FIGURE 3.2
From Harris Interactive.

Table 3.1. Americans Were Asked, "What Type of Flooring Do You Currently Have in Each of the Following Rooms?" (%)

		Carpet	Wood	Tile	Stone	Granite	Marble	Linoleum	Laminate	Vinyl
Bedroom(s)	N = 1,140	80	16	3	1	<1	<1	1	3	1
Living room	N = 1,091	65	23	6	1	<1	<1	<1	5	1
Office/den	N = 611	68	20	5	1	<1	<1	<1	2	2
Family room	N = 580	63	19	9	2	<1	<1	1	5	1
Hallways	N = 1,001	59	24	10	1	1	1	2	5	1
Dining room	N = 880	39	29	11	2	<1	1	7	6	5
Basement	N = 506	37	2	7	12	2	0	3	2	4
Bathroom(s)	N = 1,131	14	4	42	3	2	2	19	5	16
Entryway/foyer	N = 652	10	31	29	4	2	4	5	6	6
Kitchen	N = 1,139	4	14	29	2	1	<1	25	9	15
Library	N = 70	53	32	6	0	0	1	1	<1	0

Source: Harris Interactive.

Most Americans will replace their floors not because they are worn out but because they are out of style.

Ceramic

As one of the most popular and fastest-growing flooring materials in America, ceramic tile offers the benefits of durability, appearance, and ease of maintenance. Ceramic tile is widely used in both residential and commercial applications and has been around since the Romans installed the first mosaic tiles.

The U.S. ceramic tile industry has been the only manufacturing group to have both established a test method for slip resistance and led the flooring industry in providing slip-resistance data as it applies to their products (see chapter 8). Because the ceramic tile industry has had decades of experience measuring their products' slip resistance, some of the safest floor coverings are ceramic tile. However, like any other type of hard-surface flooring material, ceramic requires proper maintenance. Many individuals who experience a slip and fall on ceramic tile usually blame the surface for being

FIGURE 3.3

slippery; however, in a recent study of household floor cleaners, it was found that the problem may actually be caused by the floor-cleaning chemical and not the floor.

There are several types of ceramic tile, including glazed, unglazed, porcelain, and mosaic. The most typical type of ceramic tile is glazed, and because the wear layer or glaze is applied separately from that of the base tile, glazed tile has a consistent level of slip resistance. Several types of aggregate materials can be added to the glazing process, increasing the tiles' slip resistance. These high-traction surfaces are recommended for wet applications where slips and falls are likely.

A second type of ceramic tile is porcelain tile. What makes porcelain tile different from glazed ceramic tile is the substrate. A finer, more refined type of sand mixture is used to make porcelain tile, producing, after firing in a kiln, a harder, less porous product. Porcelain tile is often used in high-traffic areas, including airport terminals, restaurant dining rooms, and office vestibules. Because the surface of porcelain tile is very smooth, it often produces artificially high wet static coefficient of friction (SCOF) values. Another benefit to porcelain tile is that it can be produced in various textures to enhance its slip resistance. Like glazed tile, porcelain tile requires proper maintenance and should not be coated with polish.

Maintaining ceramic tile is easy, except for the grout lines. According to a Harris Interactive poll, 41% of respondents felt that ceramic tile did not present a cleaning challenge; however, 14% felt that keeping the grout lines between the tiles clean was difficult.

FIGURE 3.4
Glazed ceramic tile.

FIGURE 3.5
Porcelain tile.

FIGURE 3.6

FIGURE 3.7
Quarry tile.

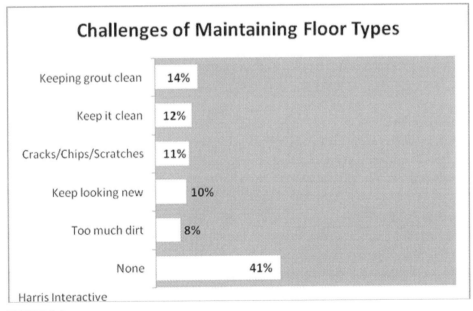

FIGURE 3.8

Since 1978, U.S. consumption of ceramic tile has grown from 549,805,000 square feet to 3,142,324,000 square feet in 2004. This works out to be a 6.93% growth rate since 1978 (Astrachan 2005).

Wood

Wood floors have been around for centuries and come in many types and sizes. Today's wood floors differ greatly from those used by our ancestors. Unlike wood floors of the past, today's wood floors often come prefinished with a urethane wear layer that not only improves the floor's durability but also enhances its appearance and ease of maintenance. Consumers no longer need to use wood-floor cleaners to maintain the appearance and cleanliness of their wood floors but can now use a wide range of products, including those designed for resilient or laminate floors.

Laminate

Another type of flooring is laminate. Laminate floors have become one of the most popular choices of flooring in America and come in a wide range of colors, patterns, and sizes. One of the most popular is wood plank. Laminate wood plank has all the richness and feel of real wood floors at a fraction of the cost. However, consumers often believe that they need to use a wood cleaner to maintain the appearance of these floors. Not so. Most wood floor cleaners contain natural oils or soaps that serve to keep the wood floor moist and to prevent the floor from drying out. These products, if applied to a laminate floor, will often leave a slippery, oily film behind, greatly reducing the floor's slip resistance. To make sure you're using the right floor-cleaning product, consult the flooring manufacturer.

Rubber/Synthetic

Rubber floors are often found in areas where heavy objects may be dropped, as in a weight room, a locker room, or an airport baggage area. Aside from daily cleaning, rubber floors require little maintenance and are extremely durable.

Choosing the Right Floor

Floors need to be carefully selected on the basis of not only their aesthetic value but also their safety. All too often, designers, architects, or even household consumers select an inappropriate flooring material, only to find this out after a slip and fall have occurred.

Table 3.2   Major Challenges of Wood, Laminate, Vinyl, and Linoleum Floors: Verbatim Comments

	Wood (N = 488)	Laminate (N = 202)	Vinyl (N = 285)	Linoleum (N = 401)
Keeping it clean	"Really getting it clean. The vacuum and Swiffer are not enough."	"It shows everything so needs everything done constantly."	"It needs to be cleaned more than once a week."	"Keeping it looking clean, it seems to stain."
Cracks/chips/scratches	"Preventing dropped items making chips into the wood."	"Keeping from scratching it."	"Dents from dropped utensils."	"As it ages we have scuffs and scratches that aren't as easy to clean."
Keep looking shiny/new	"The shine/finish has worn off and I put rugs over those spots in front of couch and lounge chairs in living room."	"Keeping it from looking dull."	"Finding a good wax that can be used to produce a shine without building up and dulling the floor."	"Keeping the shine on it."
Dust removal	"It shows dust easy, even after vacuuming I have to go over it with a Swiffer to get the dust."	"It shows dust almost everyday . . . so I dry Swiffer it almost every other day."	"Dust builds up in corners."	"Getting rid of dust."
Pet hairs/stains	"Getting rid of the cat hair."	"Keeping the dog hair off of it."	"The dogs track a lot of debris in the house, and vacuuming is a must. Leaves and mud are the biggest problem."	"With a dog it gets scratched and dirty."
Too much dirt	"Dirt gets into nail holes and small crevices and shows poorly."	"The laminate we have is light colored and the dirt is ground in and makes it look dingy."	"Light-colored vinyl and the dirt gets ground in and makes it look dingy."	"Keeping dirt out of the creases."
Stains	"Scratches and water stains."	"Keeping colored kool-aid stains off of it."	"Stains that have set in."	"Spills and dark streaks."

Source: Harris Interactive.

Table 3.3. Major Challenges of Tile, Stone, Granite, and Marble Floors: Verbatim Comments

	Tile	Stone	Granite	Marble
Keeping grout clean Stains	"The grout gets dirty more." "Stains that are hard to remove."	"Cleaning grout between the stone." "It stains easily."	n/a n/a	"Keeping grout in place and white." "Marble can be stained and hard to remove those stains."
Keeping it clean	"It is a relatively rough, industrial tile, so the Swiffer doesn't always clean as thoroughly as I would like."	"Keeping it clean."	"Keeping" it clean.	"Clean."
Keep looking new/shiny Too much dirt	"Keeping it shiny." "When we wet Swiffer, keeping the wet dirt from spreading."	"Making it look brand new." "Dirt in the crevices."	"Shine." n/a	"Maintaining shine." "Dirt that collects in cracks."
Pet hair/stains	"Tile gets dirty and dusty quickly because of the cats."	"Muddy feet and paw prints."	n/a	"When the dogs come from outside."

Source: Harris Interactive.

FIGURE 3.9
Limestone.

FIGURE 3.10
Slate.

FIGURE 3.11
Terrazo.

THE INVENTION OF THE THRESHOLD
Prior to the seventeenth century, most European floors were dirt. Only the wealthy had something other than dirt (hence the saying "dirt poor"). The most common type of flooring material used by the wealthy was slate, which, when wet, became slippery, so they spread thresh (straw) onto the surface to help keep their footing. As the wet winter months wore on, they continued to add more thresh until, when a door was opened, the thresh would blow away. So, in an effort to keep the thresh from blowing outside their homes, they would place a piece of wood in the entranceway to hold the thresh (hence "thresh hold").

FLOOR MATTING
One of the best ways to capture tracked-in moisture and soil is through the use of floor mats. However, using the right floor mat is important. Carpet mats are standard in many industries, and, if used properly, they can significantly reduce slips and falls. However, if not properly inspected or maintained, carpet mats can pose a significant trip-and-fall hazard. One of the leading causes of trip-and-fall accidents is a buckled,

folded, or migrated floor mat, usually at a building's front entrance. If not secured to the floor, most carpet mats will, over time, move or flip up on their edge when pushed up against a wall or threshold. Carpet mats are also prone to buckle when compressed under a heavy rolling load like that of a pallet jack, heavy shopping cart. or delivery dolly. Because of this, carpet mats require frequent inspections to ensure that they do not pose a safety hazard.

Floor-Matting Applications

Appropriate size is one of the most important factors affecting the proper use of floor mats in places such as lobbies, vestibules, and building entrances. Floor mats must be large enough to be walked on because small door mats are often kicked aside or stepped over, thus preventing pedestrians the opportunity to remove soil and moisture from their shoes. Mats should be of a type and construction that prevents trip hazards and should be inspected often. If edge curling occurs, tape down the edges with pressure-sensitive tape or duct tape. Curled or buckled mats present a significant safety hazard and are often the reason for trip-and-fall lawsuits.

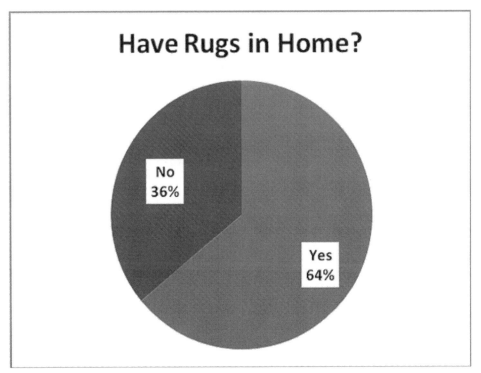

FIGURE 3.12
From Harris Interactive.

Although no formal research has been done as to how much matting is required to remove soil or moisture from pedestrian footwear, the floor mat industry has provided an estimate. The industry contends that it takes 12 feet of carpet matting to remove 80% of moisture or soil from pedestrian footwear. The National Floor Safety Institute (NFSI) is conducting a formal study on this subject that will soon be published.

It has been estimated that 85% of all soil enters a building on the feet of building occupants. Of this, at least 80% is dry soil, and the rest is oily. The dry soil can range from large particles to powderlike dust. The International Sanitary Supply Association has estimated that it costs $600 to remove one pound of soil after it has been allowed to enter an average building. This cost is primarily in labor. Entrance mats stop and contain soil and water, and removing soil from a mat can be less expensive than removing it from a building since the soil is concentrated in a localized area.

An entrance mat should do four things:

1. Stop soil and water at the door. Most if not all mats claim to do this. The most effective mats provide a combination of scraping and wiping to stop the maximum amount of contaminants.
2. Store soil and water for removal. The most effective entrance mats are designed to provide a place for soil and water to go to for storage. It is important that the storage be designed for maximum holding and ease of removal when the mat is cleaned. It is also important that the mat provide a way for the contaminants to be contained so that water cannot spread to the surrounding floor, creating a slip-and-fall hazard. Mats with flat borders allow water to seep off the edges, thus increasing one's exposure to slip and fall.
3. Minimize tracking of soil and water. A bilevel construction that provides an upper surface for walking and a lower area where soil and water are stored until removed by cleaning best accomplishes this capability. The key to performance is the depth of the construction. Mats with a depth of at least an eighth of an inch permanently molded into the mat will contain soil and water away from the feet for maximum performance. High-performance mats maintain this bilevel construction permanently, thereby trapping large quantities of moisture and dirt and minimizing the effects of soil and water being tracked further into the facility. Low-performance mats, such as those with ribbed or square-patterned face yarn, alone will not hold up under foot pressure, thus allowing more water and dirt to reattach to shoes and be tracked into facilities. High-performance mats have a reinforcing membrane that will permanently support the textile face throughout the mat's productive life, increasing product unity and extending useful product life.
4. Provide a safe surface. The mat should be slip resistant to prevent it from moving on the floor when it is walked on. In addition, any water on the mat should be

contained in a reservoir below the traffic surface to prevent slipping on the flooring surface adjacent to the mat. Some carpet mats have cleated backings that enhance their slip-resistant properties and allow moisture to dry from underneath the mat. Never place a mat over a wet floor. When a mat is placed on a wet floor, it becomes like a surfboard and can cause the mat to slide across the flooring surface and in turn create a slip hazard. Entrance mats may also be scraper mats that provide the first line of defense against soil. Scraper mats should always be used with other entrance mats that will provide wiping properties to remove fine soil and water from the feet.

Matting Types

At present, more than 50 different types of floor mats are available commercially. The types of mats selected depend on the needs of the facility, such as climate, amount of inclement weather, and floor type. A list and description of the more popular types of mats follows.

Carpet Mats (Wiper Mats)

Carpet mats are an excellent choice to protect lobbies during wet weather. They are constructed of nylon or carpet pile and are available in various sizes. These mats effectively prevent tracking of grit, fine dust, and moisture. Because of their heavy backing, carpet mats lie flat without taping. A runner is simply a long carpet mat. While most carpet mats come in standard sizes, such as 3 by 5 or 4 by 6 feet, runners are usually 3 by 10 or 3 by 12 feet in size. The backings of carpet mats are usually made of rubber or vinyl to which rubber-backed mats offer greater slip resistance.

Link Mats

Link mats effectively prevent tracking of heavy soil but pick up little moisture. They are constructed of plastic, vinyl, or rubber and can be used either inside or outside the building.

Tire Mats

Tire mats are extremely durable and effectively prevent tracking of soil. They are made from old tire casings and are low in cost. Tire mats may be used inside or outside the building.

Scraper Mats

Rubber-based mats are designed to be used to remove hard soils from pedestrian footwear and are often used outside of entranceways. Although scrapers and wiper

FIGURE 3.13

mats are used in the same application, their surface pattern and material will differ from one type to another.

Cocoa Mats

Cocoa mats are made of tough cocoa fibers. Because of their porous construction, they very effectively prevent tracking of soil and moisture.

Recessed Mats

Commonly used in front-entrance vestibules, these mats are installed into a shallow well or recess to eliminate an elevated edge. Recessed mats provide a nearly permanent

FIGURE 3.14

FIGURE 3.15

solution and are easily cleaned by simply lifting the mat from its well and replacing it. Recessed mats prevent both slips and falls as well as trips and falls.

Slip-Resistant Mats

Most often used in commercial kitchens, bars, or locations where workers are exposed to wet or contaminated floors, slip-resistant mats have an abrasive top layer and a rubber backing and come in various sizes.

Antifatigue Mats

Designed to provide comfort for workers who are on their feet for long periods of time, antifatigue mats are made of rubber or lightweight foam. Antifatigue mats are often used in commercial kitchen cooking areas and bars and behind cashier stands.

Carpet Tile

Used in entrances as well as vestibules, carpet tile offers the best of both worlds: safety and ease of maintenance. Carpet tile is used in lieu of hard-surface flooring and provides a maximum level of slip resistance. Carpet tile sales in America have dra-

FIGURE 3.16

matically risen, and it has become the industry standard for grocery and retail store vestibules as well as the health care industry.

Recessed Grated Mats

When is a mat not a mat? When it's a metal grate. Grated entrances offer the benefit of removing dry soils but do not always perform as well with migrating moisture. These "high-tech" materials require special installation to be recessed into the floor and allow dry soils to drop through the openings between the grates and collect below the walking surface.

FIGURE 3.17

FIGURE 3.18

What's the Difference between a Mat and a Rug?

Wikipedia defines a floor mat as "a generic term for a piece of fabric or flat material, generally placed on a floor or other flat surface, and serving a range of purposes," while a rug is defined as a "carpet made or cut and bound into room dimensions and loose laid." The big difference between a mat and a rug is that floor mats have a synthetic backing usually made of rubber with a tufted carpet surface and rubber edging, while carpet rugs have no backing and are constructed of woven carpet with bounded or frayed edges.

FIGURE 3.19

FLOOR CARE

Ninety-one percent of shoppers decide where to shop on the basis of whether the floor is clean (Progressive Grocer 2000).

Improper floor maintenance is one of the leading factors contributing to slips and falls. Regardless of the type of floor you have, unless you maintain it properly, its built-in slip-resistance qualities will become compromised.

FIGURE 3.20

The American Cleaning Model

Steve Spencer, facilities director for State Farm Insurance, has been a pioneer in the area of defining "clean" as it relates to floors. His company, aside from being one of the largest insurers in the country, also has approximately 30 million square feet of its own office space in which slips and falls are common. What Steve has done is to define the cleanliness of a floor according to its slip resistance. His approach relies on the idea that if a floor is truly clean, then its slip resistance should be similar to when the floor was new. If a particular flooring material had a wet SCOF of 0.6 as measured and specified by the manufacturer, then, if one wants to know if the floor is clean, one can measure the floor's slip resistance as it exists in the real world and compare it to that specified by the manufacturer. If the slip resistance is similar, then the floor is clean. For floors that have a slip resistance far below that of the specified level, one can also quantify just how dirty the floor is. Each person or company can then develop an in-house definition of cleanliness based on individual needs.

One can define cleaning as the removal of residues, where residues can be fat, protein, carbohydrate, bacteria, mineral, or a combination. Not only do these render the floor slippery, but they also supply bacteria with a nutrient source, therefore making the floor unsanitary.

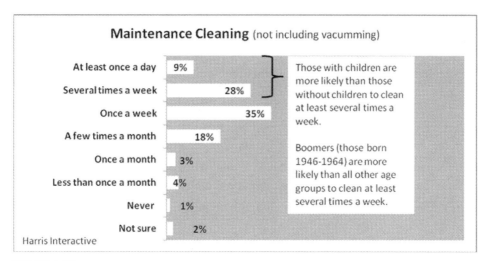

FIGURE 3.21

Professional Cleaning

> The majority of home owners have never hired professional cleaning services companies to clean their carpets or other hard surfaces.

Effects of Polymerization

Polymerization is the combined effect of wear and soil contamination. Polymerization will occur even under ideal cleaning procedures and is the leading cause of slippery conditions on commercial flooring. The real-world effects of polymerization result in reduced traction between a person's footwear and the floor.

Household Floor Cleaners

Fragrances are often oil based and may contribute to the formation of a polymerized film. It should also be noted that traditional cleaning methods have changed little, if at all, over the years.

Common consumer misconceptions exist when it comes to appearance and fragrance. Decades of advertising have convinced most Americans that a high-gloss floor is a "clean" floor and that a floor whose appearance is dull is a "dirty" floor. Furthermore, we have been led to believe that by combining our senses of both sight and smell, we can detect a clean floor. If the floor is shiny and smells good after cleaning, then it must be clean. But this raises the question of what a clean surface (floor or otherwise) smells like. Nothing, right? After all, the smell of clean is no smell. When

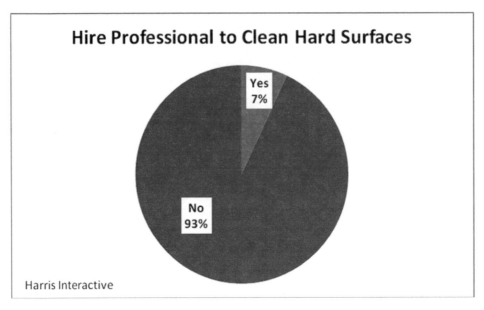

FIGURE 3.22

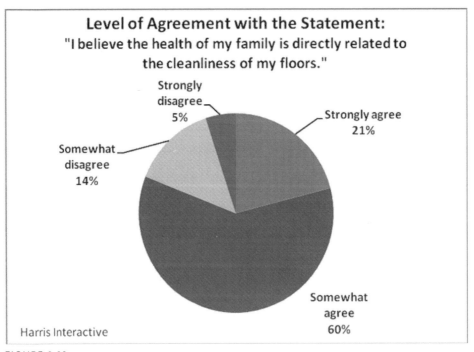

FIGURE 3.23
From Joseph M. Owens, "Technology of Floor Maintenance and Current Trends," ASTM International, pp. 3–28.

"IDEAL"

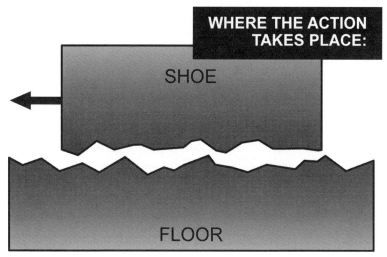

WHERE THE ACTION TAKES PLACE:

SHOE

FLOOR

FIGURE 3.24

"REAL WORLD"

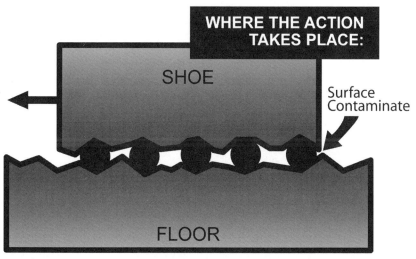

WHERE THE ACTION TAKES PLACE:

SHOE

Surface Contaminate

FLOOR

FIGURE 3.25

surgical instruments are sterilized, they do not come out of the autoclave smelling like flowers, seashores, or exotic fruits, do they? For a surface to give off a fragrance means that there must be some residual product (i.e., fragrance) remaining on the surface. And what exactly are these fragrances but perfumes of which many are oil based? So, in order to satisfy the need for gloss and fragrance, many floor care manufacturers add oil-based fragrances to their products. These chemicals can build up over time and reduce the slip resistance of the floor, thus increasing the risk of a slip and fall.

This link between gloss level and cleanliness is uniquely American. Other countries do not accept the notion that a dull floor is a dirty floor—walkways in Europe are often dull. Without a doubt, we as Americans buy in to the idea that floors should be shiny, and we spend billions of dollars making them so. Many researchers have found that shine and safety are indirectly proportional to each other and that the higher the gloss level, the more likely one is to experience a slip and fall. But the reason for this may not be what you think. Just because a floor is reflective does not mean that it has a low slip-resistance level. In many cases, it is just the opposite. Many dull surfaces do not provide high levels of slip resistance, but, unlike shiny floors, it is easier for the pedestrian to see a wet hazard on a dull floor than on a shiny surface. This is particularly true for the elderly. In fact, most nursing homes intentionally use low-gloss floors and floor finishes for this reason. Even those with the best vision find it difficult to recognize a spot of water on high-gloss floor, and the smaller the spot of water, the more hazardous it is.

> In a recent study published by *Chain Store Age* magazine, it was found that for every dollar the retail industry spends on floor care products, it spends three dollars on slips and falls.

Question: Name two ingredients in Murphy's oil soap?
Answer: Oil and soap.

Question: What is an active ingredient in Pine Sol?
Answer: Pine oil (thus the pine fragrance).

When we mop our floors with oil-based cleaners, we are actually leaving behind some type of oil or fragrance that actually makes the floors more slippery after mopping than before. So the trade-off for safety is appearance and a pleasing fragrance, a fact that most people don't give much thought to. This relationship was demonstrated several years ago by a research project conducted by the NFSI. The NFSI study concluded that

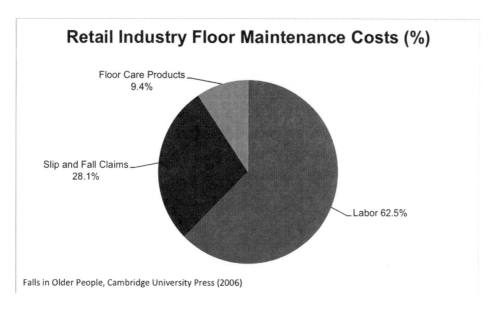

FIGURE 3.26

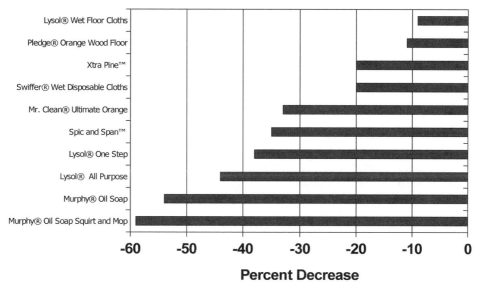

FIGURE 3.27

many of the most common household floor cleaners actually made floors more slippery after a single mopping.

Be careful of overspraying aerosol furniture and stainless cleaner polishes. These products often contain oils that, when oversprayed onto the floor, can lead to a slip-and-fall accident.

High-Traction Floor Cleaners

The use of high-traction floor cleaners and treatments has been highly effective in preventing slip-and-fall accidents. These unique products work by restoring the floor back to its original condition and without leaving a soap film. Consult your floor care supplier for more information on these traction-enhancing products. And don't confuse acid-based cleaners and safety treatments with acid etchants. All acid etchants are acid based, but not all acid-based cleaners etch floors.

One study revealed that real-world slip resistance (SCOF) will change over the time of day and that high-traction floor cleaners will provide a higher level of wet slip resistance than that of a conventional cleaning product.

Floor Maintenance Practices

Certain floor care practices also contribute to slips and falls. Among these are the following:

- Using too much or too little of a floor care product
- Using the wrong cleaner for a particular surface
- Not following label instructions on the cleaner
- Not rinsing the floor care product, if required
- Not brushing the floor thoroughly, if required
- Not having or adhering to a routine floor maintenance schedule

YOU CAN'T CLEAN A RESTAURANT FLOOR WITH A JUST A MOP AND BUCKET

You just can't! Why? Because the waxlike polymerized films that build up on kitchen floors will not be removed by mopping alone but rather require a through brushing. The problem is that brushing is hard work, time consuming, and usually done late at night when most workers (and their managers) would rather go home. Let's face it: scrubbing a dirty floor is dirty work. Decades ago, that was the way most people cleaned their floors—on their hands and knees with a hand brush and a rag.

Not to anyone's surprise, 55% of people still use the old-fashioned mop-and-bucket approach to cleaning hard-surface floors. The reason that brushing is important is due to the way soils adhere to the floor. Even the smoothest floors have microscopic pores that soils, including soap films, can penetrate and get stuck in. The most porous floors include those often used in commercial kitchen floors (i.e., quarry tile), which have absorption rates of 5% or more, making them sponges for grease, oil, and soap film. If not properly cleaned, these pores become filled and make the tile smoother and more slippery. Furthermore, most floor-cleaning chemicals are designed to bond or "emulsify" to grease and oil particles, in turn requiring rinsing. But because oil and grease are easily absorbed into the surface pores, the soap film will chemically bond to the floor, creating a waxlike polymerized film. The solution? Use an NFSI-certified floor-cleaning product, brush the surface, and rinse. This process will keep most kitchen floors from becoming polymerized and in turn create a safer walking surface.

Floor Waxes and Finishes

A finish is the final (usually permanent) protective coating applied to a resilient floor surface. The term "finish" is often used interchangeably with "polish" and "wax." Finishes are distinguishable by their solids content. Solids are the part of the finish that remain after the other ingredients have evaporated. The types of ingredients that make up the solids content are the important characteristics in determining performance.

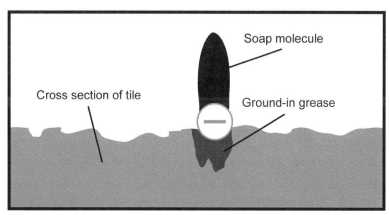

FIGURE 3.28

Cleaning Preferences
All Hard Surfaces

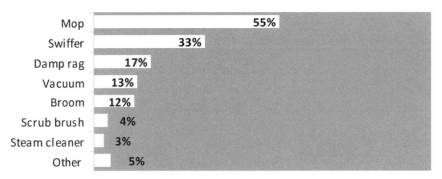

FIGURE 3.29
Harris Interactive.

Why are floors polished?

- Polishes act as a wear layer, protecting the floor and extending its life.
- Polishes improve the appearance and gloss level of the floor.
- Maintenance is easier on a polished floor surface.

Buffing/Burnishing

Burnishing is the process by which a floor's smooth and glossy appearance is enhanced. Burnishing takes place using a floor buffer or burnisher and companion burnishing pad. The frictional force between the floor pad and the floor finish causes the pad to microscopically soften the floor finish, thus improving its gloss level and appearance.

Facts about Finishes

Most commercial polishes do not contain wax but do contain as much as 100 parts per million of silicon oil as a foam-control agent. In Europe and some parts of the Pacific Rim, the normal silicone oil defoamer is replaced with 2% to 8% fatty acid in the formulation. Olefin polymers have displaced natural waxes because the synthetics are cheaper, lower in color, and more chemically and mechanically stable and provide better control of slip resistance.

Polishes formulated with a styrene-free (all-acrylic) polymer (crosses) can provide the same gloss as a styrene-containing polish, though only when more coats are applied to the floor. Applying fewer coats of higher-solids formulations can attain the same results. The application of fewer coats to attain a high film loading and high gloss reduces application time and cost.

The upper limit of gloss is called "wet gloss"—the dried film has the same gloss level as a wet floor. When wet gloss is achieved, it is often impossible to tell with the naked eye if an applied polish is wet or dry, and it makes detecting water on the floor very difficult. Because of this, wet gloss floors are a contributing factor to slips and falls. High-gloss floors may also cause disorientation or loss of depth perception for some individuals and should not be used in nursing homes, hospitals, or restrooms.

> Poor or incomplete film formation is often the root cause of many polish performance problems (Owens 2004).

The amount of floor finish applied to the floor is determined by both the composition of the mop (i.e., the type of fibers used to make a mop and how those fibers are woven together) and the amount of floor polish that is wrung out of the mop before using (Cartwright 2004). The polish film is very thin and typically measures a few thousandths of an inch in thickness. The constant exposure to pedestrian traffic and frequent cleanings combine to damage the surface of the polish and cause it to wear as time passes. Because traffic patterns are never uniform, the polish usually deteriorates irregularly. In some areas, damage is so heavy that the polish film may be completely removed from the floor. Detergent cleaning of the floor does change the surface characteristics of the polish.

> People have long associated a floor of glossy appearance with slipperiness, although, as we now know, there need be no relationship at all (Astrachan 2005).

> Research has shown that after a 30-day cycle of repeated floor scrubbing, the components normally associated with the polyethylene wax (a key ingredient of the polish) were no longer present (measurable) at the surface of the polish. Thus, the product as tested in the laboratory per the ASTM D-2047 method is not the same as that of the real world.

High concentrations of cleaner or the use of stronger, alkaline cleaners will cause film degradation to accelerate. As with spray buff mediums, a high-speed restorer is a formulation that permanently softens the polish surface to augment the abrasive smoothing and polish film surface removal processes. High-speed burnishing equipment completely removes most restorer formulations. As in the case of spray buffing, burnishing requires specially designed formulations for best results. The most important features are exceptional durability and slip resistance. Without a doubt, a softer polish is easier to abrade, and it will respond readily to the burnishing process. However, the film formed by such a formulation will be extremely prone to deep scuffing, marking, and soiling. Most important, though, a softer polish will not last as long as a polish based on a "harder," more durable polish polymer (Tysak 2004).

> The damp-mopping procedure leaves the built-in slip-resistance qualities of polishes essentially unchanged.

"Traffic on polishes, when not accompanied by cleaning maintenance, lowers the static coefficient of friction of polishes, and reduces the slip resistance of the film. Ingrained soil in the film is the cause of this deterioration in slip resistance. Scuffing, loss of gloss (micro-scratches), and black marking have no measurable effect. Spray buffing temporarily reduces polish micro-scratches and high speed burnishing temporarily increases polish micro-scratches" (Owens 2004).

By combining a high-quality, traction-enhanced floor polish with a proper buffing regime, you can improve a floor's safety. Contrary to what many people believe, a shiny floor is not always unsafe. High-traction floor finishes that act to prevent wet-surface hydroplaning should be used in high-traffic or high-risk areas.

Paints and Coatings

One of the most common materials used to prevent concrete from staining is to apply an epoxy-based coating to its surface. Common applications include parking garages, warehouse floors, and manufacturing lines. The primary reason for applying such coatings is to provide a sealant to the floor to protect the concrete surface from chemical damage or water migration from the subfloor. Epoxy-based coatings provide a visual enhancement and ease of maintenance. Many epoxy-based paint manufacturers will recommend that a mineral aggregate additive be applied to the floor. Such additives come in a wide range of grit sizes and can dramatically increase the slip resistance of the floor. Unfortunately, these additives can wear off quickly, leaving a slippery surface behind. Once the aggregate has worn off an epoxy type surface, it can be reapplied using an additional layer of paint and additive. However, it is a complex and costly project, often requiring the removal of the underlying epoxy paint.

Textured coatings, including "poured floors," are similar to epoxies but are applied using a textured paint roller or trowel. The rougher the surface of the roller, the greater the texture of the surface and the greater the level of slip resistance. Textured coatings are most often used on industrial floors, exterior walkways, and areas where oil and grease are often present. These products are so popular that even the military uses them on the surfaces of its ships.

Grit Diameter (inches)	Grit Diameter (microns)	Mesh Size	Examples
0.008–0.006	254–145	60–80	Restaurants / food preparation
0.014–0.008	356–254	40–60	Food processing
0.027	686	20	Manufacturing workstations
0.073–0.053	2210–1346	8–12	Vehicle ramps

BUILDING SERVICE CONTRACTORS

Inadequate Housekeeper Training

Across industries, employees may lack the training needed to perform the proper floor maintenance procedures. There are several reasons this may be the case:

- *Poor Orientation.* New employees may not be properly trained when they begin working for the firm. Often new employees are sent immediately into their jobs, leaving overextended managers to provide initial training. Some organizations require new employees to complete a training program and checklists as a part of their initial orientation. Floor safety training can be performed either person to person (i.e., on-the-job training) or in a more structured classroom setting. Written procedures can play an important role. However, what is most important is the degree of consistency in communicating the actual floor safety procedures and policies. Written safety policies and procedures are ineffective if the employees do not understand or follow them.
- *Uneven Skill Level of Employees.* Employees come to an organization with varying backgrounds. Management may assume that employees are familiar with basic maintenance procedures and with their responsibility to protect their own safety and that of their coworkers and guests. Just because an employee came from another company within your industry does not mean that he or she was properly trained.
- *High Turnover.* High turnover in certain industries (retail, food service, and so on) often makes it difficult to provide a consistent level of training. Most large retailers outsource their floor maintenance, usually through a building service contractor (BSC). However, from 1999 to 2006, the percentage being outsourced remained at a constant level of 77% (International Facility Management Association 2006). When

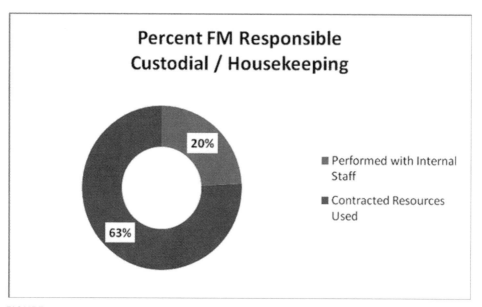

FIGURE 3.30
From Facility Management Practices: Maintenance Operations Responsibilities (Research Report #16).

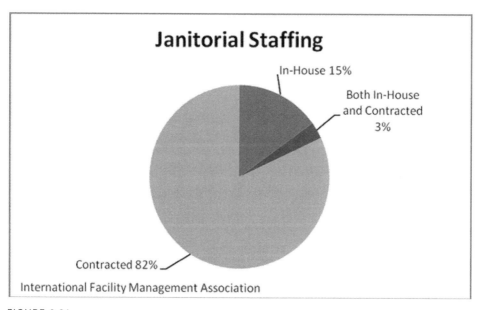

FIGURE 3.31
From Operations and Maintenance Benchmarks: Janitorial Staffing (Research Report #26, p. 29).

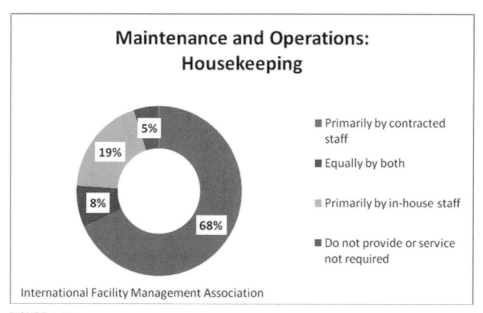

FIGURE 3.32
From Outsourcing Management Practices: Maintenance and Operations (Research Report #27, p. 6).

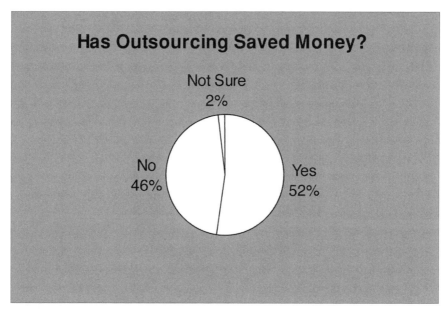

FIGURE 3.33
From Outsourcing Management Practices: Has Outsourcing Saved Money? (Research Report #27, p. 21).

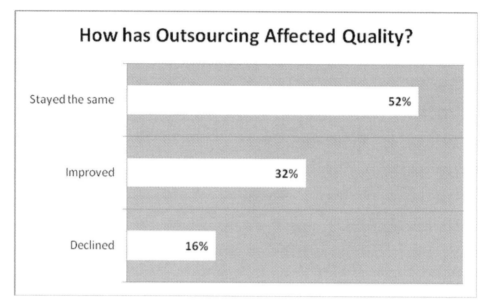

FIGURE 3.34
From Outsourcing Management Practices: How Has Outsourcing Affected Quality? (Research Report #27, p. 21).

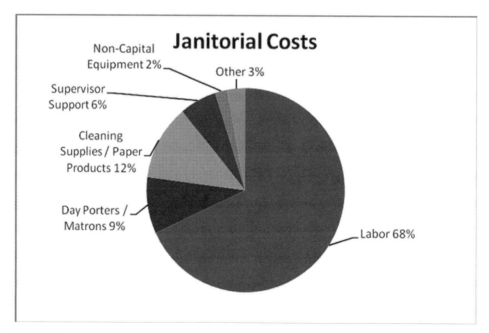

FIGURE 3.35
From Operations and Maintenance Benchmarks: Janitorial Costs (Research Report #26, p. 29).

those who have outsourced their housekeeping functions were asked if they saved money, only half said yes. And when asked about the quality of work provided by outsourcing, 52% of respondents saw no improvement in quality.

"Seventy-two percent of small facilities have responsibility for administrative service areas with a majority performing with internal staff. Only 44% of the larger facilities have this responsibility and only about one-fourth perform this work with internal staff" (International Facility Management Association, 1996).

How Often Are Floors Cleaned?

According to the International Facility Management Association (2006), only 7% of BSCs will sweep and mop commercial VCT on a daily basis and only 10% weekly. This is to say not that such floors are rarely cleaned but rather that they are not cleaned by the BSC.

Maintenance and Cleaning
- Seven in 10 home owners clean their floors at least once a week. About the same percentage vacuum their carpets at least once a week. However, when it comes to deep cleaning their floors (polishing wood or marble, cleaning carpets, and so on), U.S. home owners do it with less frequency—nearly half deep clean their carpets, and 45% deep clean other hard surfaces at least once every six months.

Health and Allergies
- Eight in 10 U.S. home owners see a direct relationship between the health of their families and the cleanliness of their floors.
- A third of U.S. home owners say that someone in their household suffers from some form of indoor allergies. Half these home owners believe that the type of flooring used causes or aggravates allergies.
- Wood floors are seen to be the most effective in controlling allergies and most effective for improving indoor air quality. Carpets are seen to be the least effective in minimizing conditions that aggravate allergies.

How Does a Property Owner Ensure
That Their Floors Are Being Properly Maintained?

It all begins with employee training. Workers need to be properly trained as to how to inspect a floor for both cleanliness and safety. Simply walking from the front of the building to the back does not constitute an inspection. Floor inspections are designed to find things that are out of place, such as spilled merchandise, improperly cleaned floors, and potential slip, trip, and fall hazards, many of which take a trained eye to recognize. Often a hazardous condition may exist but is invisible to the naked eye.

Table 3.4. How Often Various Cleaning Tasks Are Performed by Janitorial Staff

	More Than Once a Day	Once a Day	Semiweekly	Weekly	Biweekly	Monthly	Quarterly	Annually	Biannually	As Required	Not Performed
Sweep/mop tile/composition flooring	7%	69%	7%	10%	1%	2%	1%	0%	0%	2%	1%
Buff tile/composition flooring	1%	7%	5%	22%	4%	21%	17%	5%	2%	13%	3%
Wax/polish tile/composition flooring	0%	2%	2%	5%	1%	15%	28%	15%	11%	16%	5%

Source: International Facility Management Association (2005).

One example is a spilled lubricant, such as motor oil, silicone spray, or aerosol air freshener. These products, when deposited onto the floor or improperly cleaned, may leave a slippery residue that is nearly impossible to see, especially on a high-gloss floor. Frequent inspections, including walkway audits, can serve as a means of early detection. For too many property owners, the first sign of a hazardous floor is when a person slips and falls.

Walkway inspections are everyone's job and should be part of every company's safety manual. Only continuous floor sweeps by a store's personnel rather than timed inspections or sweep logs (which often do not provide detailed information regarding the condition of the floor) provide the kind of vigilance it takes to prevent slips, trips, and falls. The problem with hourly sweeps or timed sweep logs is that they tend to be seen by employees as "someone else's job." A manager, an assistant manager, or some other delegated employee is charged with the responsibility of conducting the floor sweep, and this leads other employees to adopt an "it's-not-my-job" attitude. Floor safety is "someone else's responsibility." Although hourly floor sweeps are the industry standard for most retailers, continuous inspections by all employees are the better choice.

Those who say that a policy stating that "safety is everyone's responsibility" means that safety is no one's responsibility are incorrect. This assessment is often made by individuals unfamiliar with the safety process. The safety process is most effective when it's shared by all workers across all levels of management and authority. The view that "safety is not my responsibility but someone else's" is the logical alternative and produces negative safety behavior often resulting in one worker blaming another when an accident occurs.

The Gleason Agency has created an automated floor sweep that relies on technology to do the work and takes the guesswork out of walkway inspections.

"Reasonable care helps demonstrate that the business owner was not negligent for the accident. Reasonable care includes a written policy, employee training, enforcement of policies, and good floor monitoring procedures," says the Gleason Agency's chief executive officer, Robert A. Gleason. A sweep log originally involved keeping a record of when someone would actually go around the store with a broom, removing hazards, although a broom is rarely involved today. "The problem, though, is that these things have lost credibility over the years," Gleason points out, "because everybody in claims and any attorney who's ever tried to defend one of these can tell you stories where the whole 24 hours of a sweep log was done in the same hand writing with the same pen. You can cheat on these things real easy just by sitting in the office banging it out. So sweep logs lost a lot of credibility in defending claims."

And too many lost claims drive insurance premiums up. When Grocers' Insurance representatives go out to train a customer on preventing slips and falls, they point out that making up a $5,000 increase in premiums due to poor performance requires sales of $500,000 at a profit margin of 1%.

The Gleason Agency has advanced the state of the sweep log with its Gleason Electronic Slip/Fall Prevention (ESP) floor-monitoring system currently used in more than 100 supermarkets. It is composed of two handheld data retrievers that collect store information, one exception card that identifies hazards and documents responses, and one transmitter that electronically relays information to Gleason's risk-management department for that of the retailer. A customized, detailed floor plan identifies strategic placement of between 17 and 22 ID markers in a typical store. As the employee—usually the manager—walks from location to location according to a predetermined route, the data collector is touched to the location button, and the date, time, location, and name of the associate are recorded. If a slip-and-fall condition is found, the employee calls for assistance, waits for the cleanup associate, and documents that a spill has been found and addressed by touching the data retriever to the proper exception card button.

Data are downloaded once a week or in the event of an incident, and management reports for each store are run weekly. Store-by-store comparisons are done weekly. Gleason says that the system costs less than $3,000 per store.

Joe Alford, director of loss prevention at Harvey's Supermarkets, a 45-store operator headquartered in Nashville, Georgia, says that since mid-November his company has been using Gleason ESP in the 16 stores that accounted for 88% of dollar losses in customer accidents. "We're real satisfied with the return on investment," Alford says. "Truthfully, the store managers would rather not do the walks. Initially, that was just one more thing to do, but now they're seeing that when your losses are probably down 60%, it's putting money in their pockets because they are heavily bonused. We have implemented a charge-back system where, if they do 75% of the ESP walk and they have an accident, they're only charged $3,000 for that accident. Otherwise, they're charged $6,000."

Alford says that the system forces managers to get out of the office and walk the floor, where they can interact with the customer and spot merchandising opportunities. "It's not rocket science," he says, "but it works."

"The accident that never happens is the one that's least expensive." (Ray Bucci, Stop & Shop)

THE BLAME GAME

It's Not My Fault

The single biggest problem in the battle to prevent slips and falls is created by the refusal of those key industries that produce flooring materials, floor care products, and coatings to accept their share of the problem.

Let's use the typical scenario in which a grocery store chain, faced with dozens of slip-and-fall claims each year, goes out to seek a solution. The grocery store usually starts with its insurance provider, who sees the problem as a training issue and recommends an informative slip-and-fall prevention program. Although insightful, such programs rarely produce tangible solutions. The grocer then raises the issue with its vendors, starting with its current flooring suppliers. Most manufacturers of flooring materials are of little help, and some even consider the discussion of slips and falls a dangerous subject. Why? Liability. Remember, as long as the manufacturers are not being sued for slips and falls, they have little interest in becoming a codefendant in your lawsuit.

FIGURE 3.36

When resilient floor-covering manufacturers are asked, "Why is the VCT they supplied slippery?," they will often respond by saying that the reason people are slipping and falling in your store is not the fault of the VCT material but rather the finish you had applied to it and suggest that you contact the finish manufacturer. When the grocer calls its floor finish supplier (usually a janitorial supply distributor) asking the same question, the most frequent answer is, "It's not our fault! Our products all meet the 'current industry standards for slip resistance,'" and that you need to speak with your floor-cleaning contractor, claiming "They may not be cleaning your floors properly." Finally, the grocer contacts its floor-cleaning contactor, again asking the same question, only to hear them say, "It's not our fault! We use the products you supplied us (or authorized us to use), and since we are not on-site during the day, we suggest that you speak to your in-house day crew—after all, that's when people are slipping and falling!"

In the end, everyone who is in a position to assist the grocer in solving its slip-and-fall problem choose to pass the buck and leave the grocer (and its insurance carrier) holding the bag. Shameful—but that's the reality.

You may ask yourself, why? Why would the flooring industry not want to take a proactive approach in preventing slips and falls? Why would the floor maintenance and finish manufacturers not directly engage in open discussions regarding floor safety? Why don't the building service contractors (BSCs) educate themselves and in turn serve as a resource to their clients? In a word: liability. The fear of being dragged into "someone else's lawsuit" has created a wall that product manufacturers and BSCs have consciously chosen not to scale. This makes manufacturers or BSCs not the bad guys but rather spectators to a high-stakes game of litigation that they would rather watch from the sidelines than participate in.

REFERENCES

Astrachan, Eric. Tile USA. E-mail referencing information from the International Society of Ergonomics Conference on Slip and Fall Issues, 2005.

Cartwright, Brian T. "The Interaction and Performance of Commercial and Experimental Fluorosurfactants and Commercial Floor Polish." *Technology of Floor Maintenance and Current Trends.* ASTM International, 2004, 29–34.

International Facility Management Association. Facility Management Practices. Research Report #16. Houston: International Facility Management Association, 1996.

———. Operations and Maintenance Benchmarks: Janitorial Practices Chart. Research Report #26. Houston: International Facility Management Association, 2005.

———. An Inside Look at FM Outsourcing. Research Report #27. Houston: International Facility Management Association, 2006.

Owens, Joseph M. "It's What's on the Inside That Counts—The Chemistry of Floor Polishes." *Technology of Floor Maintenance and Current Trends*. ASTM International, 2004, 3–28.

Progressive Grocer. Advertisement, 2000.

Tysak, Theodore. "Polish Maintenance for Fun and Profit." *Technology of Floor Maintenance and Current Trends*. ASTM International, 2004, 61–78.

4

Footwear

Shoes come in all types and styles, and the consumer usually assumes that many are slip resistant. This assumption is often wrong and contributes to thousands of slip-and-fall injuries each year. Street shoes are not safety shoes and should not be worn in areas where workers are exposed to wet or oily hazards. Footwear outsoles should be evaluated for their slip-resistant qualities. What exactly constitutes a slip-resistant sole is a combination of structural characteristics, including sole material, sole pattern, and wear performance across various environmental conditions. Just because a certain type of shoe outsole is safe for use on a construction site doesn't mean that it is safe in a kitchen in a fast-food restaurant. Wearing the right shoe can make a big difference in preventing slips and falls.

When selecting the right type of shoe for a given application, it is important to consider the following.

OUTSOLE MATERIAL

Most consumer footwear outsoles are made from a combination of compounds, including vinyl plastics, natural or synthetic rubbers, and fillers. Generally, rubber compounds, specifically Nitrile® rubber, offer superior slip resistance to vinyl polyvinyl chloride (PVC) compounds. Although PVC-soled shoes wear longer than rubber-soled shoes, rubber-soled shoes tend to conform better to a floor's natural texture than PVC and therefore provide better slip resistance. Most outsoles are made of combinations of materials often referred to as compounds. Modern outsoles may consist of five or more compounds. PVC blends make up the largest segment of sole materials sold worldwide, while leather represents only 7% of the outsole market. What is unique about leather as opposed to synthetic materials is its reputation for being slippery. Leather, when new, can be very slippery, especially on smooth, dry surfaces. However, leather becomes less slippery when broken in. In addition, because of the hydrophilic (or water-absorptive) quality of leather, the wearer often finds that leather-soled shoes may be less slippery on wet surfaces.

A CANADIAN STUDY

According to the Expert Panel on Reducing the Number of Slips and Falls in Canadian School District Facilities, "In Britain, footwear plays an important role in preventing slip incidents. Establishing a 'sensible shoe' policy (for example, flat shoes that enclose the whole foot, not sandals or sling-back shoes) has been shown to make a significant impact on reducing slip and trip injuries. Ideally, such a policy should cover all staff, including cleaning and catering staff and pupils."

The Canadian Centre for Occupational Health and Safety (1999a) offers a different type of reference table. Aside from the column headed "slipping," table 4.2 addresses more "resistance to absorption" and/or "resistance to damage."

SOLE PATTERN AND TEXTURE

The pattern or "tread" of a sole is as important as its material composition. The recommended tread will possess a pattern extending over the whole sole and heel area. Look for a minimum width of 2 mm and depth of 4 mm to appropriately channel

Table 4.1.   Sole Material Rated against a Walking Surface

Sole Material	Tile (dry)	Tile (wet)	Wood (dry)	Concrete (dry)	Concrete (wet)
Neoprene	R	R	R	R	R
Crepe	NR	NR	NR	R	R
Leather	R	R	R	NR	R
Soft rubber	R	NR	NR	R	NR
Hard rubber	NR	NR	R	NR	R

Note: R = recommended; NR = not recommended.
Source: Mekeel (2003).

Table 4.2.   Material Qualities

Sole Material	Abrasion	Chemical	Cement	Slipping	Water	Oil
Vibram	X	G	X	X	X	G
Leather	F	F	G	G	P	F
Neoprene	X	X	X	G	X	X
Neo crepe	G	F	G	X	G	G
Rubber (vulcanized)	X	G	X	G	X	G
Rubber (Nitrile)	X	X	X	X	X	X
Polyurethane	X	X	X	G	X	X

Note: X = excellent; G = good; F = fair; P = not recommended.
Source: Canadian Centre for Occupational Health and Safety (1999b).

Dissection of A Slip-Resistant Out Sole

Interlocking pattern extends to the edge of sole and is rounded for adequate channeling of liquids.

Level toe spring to maximise forward slip resistance while minimizing trip hazards.

2 mm wide by 4 mm deep channels to force liquids away from the sole, reducing hydroplaning.

Rubber compound for flexibility and walking surface conformity.

Flat arch for maximum contact area.

Flat, even heel made of patterned material.

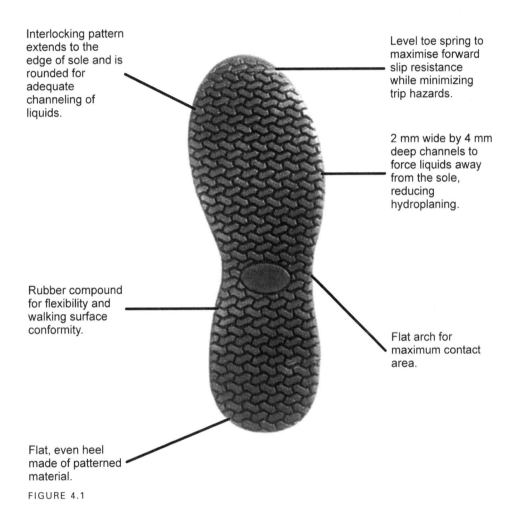

FIGURE 4.1

liquids from under the sole. Shoes that possess large, unpatterned areas or circular patterns may trap liquids under the sole, causing the wearer to hydroplane, and should therefore be avoided.

Over time, wear and tear will affect most outsole materials. Irregular heel wear, sole pattern wear, and contamination will greatly affect a shoe's performance. When selecting appropriate slip-resistant footwear for an employee footwear program, it is important to examine the surface on which the shoes are to be worn. High levels of animal fat, grease, or petroleum oils can dramatically affect the slip resistance of footwear outsoles. Terms like "oil resistant" do not mean slip resistant. An oil-resistant outsole material will repel oils, thus preventing their absorption, leading to the eventual breakdown of the material.

Temperature also plays a role. Many PVC outsoles will become rigid and crack in low-temperature environments. This rigidity not only prevents the foot from flexing properly but also prevents appropriate surface contact and traction. Avoid the use of soles that contain closed circular patterns which do not provide for channeling of fluids from the center of the sole to the edge. Such patterns trap fluids and serve to enhance the hydroplaning effect created by walking on wet surfaces.

HEELS

Don't confuse outsole materials with heel materials. Heel materials, especially for dress shoes, are often different from that of the outsoles. Heel materials may vary widely but are generally made of materials like PVC or rubber. Since many slips first occur at heel contact, it is important to select shoes with appropriate heels. High-heeled shoes should obviously be avoided in the workplace because of the low contact area of the heel. In addition, as heel area is reduced, the force on the heel area becomes greatly increased. For example, a quarter-inch stiletto heel worn by a 120-pound woman will exert more than 1,500 pounds per square inch. That's more pounds per square inch than is exerted by an elephant. High heels also force the foot to contact the floor differently. They exert more force on the toes and the ball of the foot, making it difficult to negotiate smoother walking surfaces.

When selecting shoes for wet or oily walking surfaces, make sure that the sole pattern allows for channeling of fluids from beneath the sole. A good rule of thumb when selecting appropriate footwear is that the larger and deeper the pattern, the greater the slip resistance for outdoor applications (construction, hunting, and so on) The smaller and shallower the pattern, the greater the adhesion on smooth surfaces. Avoid outdoor footwear when working indoors since larger outsole patterns reduce surface contact area on smooth floors, making them less slip resistant.

HIGH HEELS

Obviously, property owners cannot make their floors safe for all types of footwear. One example is high-heeled or platform shoes. Pedestrians must use good judgment and select appropriate footwear. Although high-heeled shoes have been the height of fashion for centuries, with them has come an elevated level of foot-related injuries. The trade-off is safety for fashion.

The fact that more women are likely to experience a slip-and-fall injury than men is often attributed to the fact that women wear high-heeled shoes and that such footwear must be the reason. The fact is that high heels do not significantly contribute to slips and falls. However, the wearing of high heels causes several deleterious effects. The lumbar flexion angle decreases significantly as heel height increases. This not only creates a more unstable posture because of the increase in the height of the center of body mass (i.e., the upper body becomes more heavy as shown by the direct three-dimensional motion analysis of the center of body mass) but also creates additional compressive forces in the lower lumbar spine because of the change in the lumbar lordosis. In addition, there is a compensatory increase in erector spinae activity to maintain the abnormal posture. All these effects can significantly increase discomfort and fatigue levels in those wearing high heels, especially at work, which explains the frequent complaints of low back pain. Although only young females were tested, these results can be expected to hold true in older subjects as well. Thus, workers wearing high heels should be strongly discouraged (Lee, Jeong, and Freivalds 2001).

FASHION VERSUS SAFETY—TAKE YOUR PICK!

The same can be said for children's footwear. About a decade ago, the first line of roller-wheeled shoes hit the market and became a multi-billion-dollar industry almost exclusively aimed at children under the age of 12. Although roller-blade shoes are fun to wear, they too bring an elevated risk of injury. The problem is so pronounced that many retailers have banned wearing them in their stores.

FLIP-FLOPS AND SANDALS

While the exact origin of flip-flops is unknown, they are thought to have originated in Japan since the flip-flop (or a close form of them) is the traditional shoe for a woman in Japan. Others have speculated that flip-flops were created in Jerusalem around the time of Christ or even before that. Researchers have found pictures of flip-flops in ancient Egyptian murals dating back to about 4000 B.C. A pair made of papyrus leaves discovered in Europe are thought to be about 1,500 years old.

Today, millions of people wear flip-flops. In some developing countries, the flip-flop is the only available shoe because of its convenience and low cost. However, in many other countries, specifically the United States, flip-flops are more of a fashion statement. Available in hundreds of styles and colors, most young American females own several pairs and will spend anywhere from $2.00 to $200 for them. They have become so socially acceptable that they have even infiltrated the White House: not too long ago, there was a scandal when the Northwestern University's national champion-ship women's lacrosse team went to visit the White House in flip-flops. Flip-flops are an ancient trend the popularity of which has caught fire and are estimated to be a $20 billion industry worldwide.

Flip-flops can cause serious problems for the wearer's feet, ankles, knees, hips, and back and are often associated with slip-and-fall incidents. The soles of modern flip-flops are made by blowing a combination of synthetic rubber materials such as styrene, isoprene polymer, benzene, and ethylene into a mold to form a lightweight foam material. All these chemical mixtures are known as being harmful to the environ-ment. Flip-flops are usually made from a material that wears quickly and becomes very smooth. When a wear pattern forms, usually on the heel, the wet slip resistance drops to dangerous levels and acts more like a surfboard than a shoe.

According to Dave Ludwin, general liability risk control director of CNA Insurance Company, "Slips and falls of patrons are the most common general liability claim in-surers face from bars, restaurants, and taverns," and he believes that many businesses may soon prohibit their customers from wearing footwear such as flip-flops.

One example of a nightclub that has already begun banning flip-flops is the Osprey Nightclub in Manasquan, New Jersey. In May 2009, the bar posted handwritten signs about the dress-code change. Osprey owner Diane Bisogni said that she's aware that the ban may be unpopular with some of her patrons but that it was a matter of safety after a flip-flop–wearing customer fell down three steps when someone behind her stepped on the back of her sandal.

"It was a decision we had to make for everyone's safety and everyone's well-being," Bisogni said. Bisogni said that doesn't mean that all sandals are banned. Open-toed shoes are acceptable as long as they have a strap around the ankle or the instep, she said. "I know we're at the shore, but I'm not the only club that has a dress code," Bi-sogni said, adding that the idea was suggested by her insurance company last year but that she resisted the change until the incident occurred.

In my experience as an expert witness, I have seen a number of slip-and-fall cases where individuals wearing flip-flops have experienced a violent slip and fall, usually as they were walking across a smooth, wet surface. Victims usually describe the slip-and-fall event as happening so quickly that they had no time to recover and often don't recall exactly what happened. Unlike other types of footwear, sandals and flip-flops elevate both the slip-and-fall and the trip-and-fall risk at the same time. Wearers can easily catch the exposed front sole on a loose carpet mat or rug just as easily as they can hydroplane on a smooth, wet floor.

Flip-flops can also present an elevated slip-and-fall risk when they come in contact with dry surfaces, such as carpeting, especially carpeted stairs. Business owners must be vigilant to keep their floors dry, specifically during the summer months, when sandals and flips-flops are most likely to be worn. Although most wearers of flip-flops are conscious of the slip/trip risks, they often find themselves the victim of a violent slip-and-fall event. It is not unusual for persons wearing flip-flops to "slip and flop." Whatever benefit flip-flops provide by way of comfort they lose in safety. Without a doubt, foam-soled flip-flops are one of the most dangerous types of footwear on the market.

ENVIRONMENTAL FACTORS

Certain shoes will wear better in certain conditions. Wedge or flat soles are preferred for indoor occupational use, such as in hospital, food service, and hospitality environments. In addition, make sure to keep the outsoles clean of accumulated debris or contaminants. Soil buildup can restrict the channeling of fluids away from the sole and contribute to a reduction in slip resistance. Property owners cannot always predict the types of shoes their invited guests will wear and should assume the worst-case scenario when selecting the type of flooring material used in their facilities, especially the entranceways. A growing number of retailers have prohibited the wearing of roller-heeled footwear like that worn by children. Such footwear elevates the risk of a slip and fall and should be worn with extreme caution.

> It has been estimated that 60% of people wear the wrong-size shoe.

FIT AND SIZE

Shoes should fit the wearer snugly but not tightly. Improper fit can lead to a slip-and-fall accident.

Only the wearer knows when the fit is proper. The following tips may assist in the fitting of footwear at the time of purchase:

- Appropriate hosiery/socks for the environment in which the footwear is going to be worn should be used when trying on footwear.
- Any extra sock liners, arch supports, orthotics, or insoles must be in the footwear when trying on shoes. If an inserted insole is added to the shoes and extends under the protective toe cap, impact and compression resistance may be reduced.
- Walk in the shoe to ensure the proper fit.
- Have your foot sized annually. Feet should be measured regularly because foot sizes change, especially with age. Both feet should be measured, and footwear should fit the larger foot. The ball (the area of protrusion at the base of the big toe) of the foot should fit well into the widest part of the shoe.

- Don't expect shoes to stretch to fit. If they don't fit initially, they won't fit later.
- Judge shoes by fit, not the size marked. Sizes vary by brand and style.
- Select a shoe that conforms to the shape of the foot. Normally, feet swell during the day. Footwear should be appropriately sized to accommodate this slightly larger size.

REFERENCES

Canadian Centre for Occupational Health and Safety. Prevention of Slips, Trips and Falls. http://www.ccohs.ca/oshanswers/safety_haz/falls.html, June 10, 1999a.

———. "PPE—Safety Footwear." *Safety Infogram*, 1999b, 1.

Chang-Min Lee, Eun-Hee Jeong, and Andris Freivalds. "Biomechanical Effects of Wearing High-Heeled Shoes." *International Journal of Industrial Ergonomics* (2001).

Mekeel, D. "Slip, Trip and Fall Prevention in Schools: Reducing Workers' Compensation and General Liability Claims Costs." *Safety Sense*, June 2003. http://www.psbait.org/Newsletters/SafetySense/inss0603.html (accessed February 22, 2004).

5

Bathtubs and Showers

According to the Home Safety Council (2004), "Over 25,000 accidents occur in or around bathtubs and that no room poses a bigger threat to safety for the elderly than the bathroom. Slip and fall accidents when entering and exiting the bathtub or shower contribute to nearly 25% of hospital admissions for people age 65 and older and the most common result is a hip fracture."

STANDARDS

Bathroom slips and falls represent the leading cause of guest injuries for the hospitality industry, most of which occur getting in or out of a guest bathtub. The problem is two-fold. First, many hotels do not provide the proper grab bars that individuals can hold on to as they get in and out of the tub. The American Society of Mechanical Engineers (ASME) A112.19.4M-1994, section 4.2 of the standard, requires that "grab bars are to be installed in the critical support area in accordance with 5.1.1–5.2," whereby the critical support area is defined under section 2.4 of the standard as "that portion of the back, service, or non-service wall in which support would most likely be beneficial in four different bathing areas." Section 5.1.1 of the standard requires that "[a] horizontal grab bar or bars shall be installed on the back-wall critical support area with a total minimum length equal to 30% of the horizontal length of the critical support area." Section 5.1.2 further requires that "[a] horizontal or vertical grab bar or bars shall be installed in the critical support area on either the service wall or nonservice wall. The horizontal bar or bars shall provide a minimum grippable length of 9 in. (230 mm) within the critical support area. The vertical bar shall provide a minimum grippable length of 6 in. (152 mm) and shall be installed in the tub entrance area."

Second, many bathtub manufacturers do not safety treat the entire bottom surface of their bathtubs, thus leaving a two- to three-inch ring of untreated tub bottom. Although this untreated portion of the tub does not play a significant role for those who use the tub to bathe in, it significantly increases the risk of a slip and fall for those who use the tub as a shower. The current ASTM F-462 standard requires that bathtubs have a minimum slip-resistance value of 0.04, measured under wet, soapy conditions. My view, like that of other safety experts including Robert L. Kohr (1991), is that the slip resistance of bathtubs should be no less that 0.1. However, such a value is below that required for the average bather to remain safe.

Section 3.3.6 of the ASME A112.19.4M-1994 standard requires that "the bathing surface of a bathtub shall be treated in such a manner that it shall comply with the ASTM F-462. Treatment shall start 2 in. measured from all side and wall radii and 3 in. measured from the centerline of the drain and from the compound corner radii."

Third, many hotels use the wrong type of flooring material in the bathroom. Products like polished granite or marble offer a high level of aesthetic value but do so at the expense of safety. Floors that do not meet a minimum wet static coefficient of friction value of 0.6 per ANSI/NFSI B101.1-2009 are inappropriate for bathroom floors and should be avoided.

Since most hotel guests use the bath as a shower stall, it is important to provide vertically mounted service wall (sidewall) grab bars so that guests exiting the tub can do so safely. The failure to provide sidewall grab bars is one of the leading causes of hotel bathtub slips and falls.

BATHROOM SLIP-AND-FALL EXPOSURES

Manning and Smith's Tech Topics (Manning and Smith Insurance 2000)

In our history, over the past 23 years, we have found that 41% of all general liability and workers' compensation claims are a result of slips and falls. Of these, 32%, or about one-third, are due to moisture on bathroom floors. Many include broken bones, back and head injuries, and even fatalities.

The bathroom is a very likely area for slip-and-fall accidents. The area is wet, soapy, and slippery. Slips and falls in bathtubs and showers are the most common types of accidents. Proper slip resistance of bathtub surfaces is the key to preventing these accidents in the first place. It is also important to furnish handrails or sturdy grab bars that remain fixed when grabbed for support.

The floor surface is the single most important factor that can contribute to a slip-and-fall accident. If an existing floor surface is relatively slippery, then the ideal remedy is to replace the floor surface with one that has more slip-resistant properties. However, this may not always be possible, and there are other remedies. There are a multitude of slip-resistant floor coatings and mats commercially available that are designed to solve many problems of the existing floor surface.

Slip-Resistant Coatings

Slip-resistant coatings can be applied to an existing floor surface and are designed to improve the coefficient of friction. There are a variety of coatings commercially available. Each has different characteristics and benefits. It should be noted that slip-resistant coatings are subject to wear, as they are only a surface coating. They require periodic dressing (maintenance) and eventually reapplication to maintain the desired level of slip resistance. Rubber or plastic coatings include urethane polymers, epoxy, acrylic, and vinyl ester resins. These coatings adhere to the existing substrate or floor surface and provide a higher coefficient of friction than the existing surface. These coatings can be applied by brush or by spray gun. Water-based floor coatings contain no volatile organic solvents. These are not flammable, and they are environmentally safer. Usually an aggregate such as aluminum oxide is included in the base material. These coatings will generally retain their nonskid properties in wet or oily conditions. They are also good for do-it-yourself applications. Since the water-based coatings contain no organic solvents, they may be used when limited ventilation or explosion hazard exists.

Slip-Resistant Treatments

Slip-resistant treatments help prevent dangerous falls simply by applying a low-maintenance liquid that microscopically etches an invisible tread on virtually any ceramic, porcelain, quarry tile, marble, granite, terrazzo or cement surface, making it slip resistant for a minimum of two years (five years for bathtub applications).

REFERENCES

Home Safety Council. *The State of Home Safety in America: Facts about Unintentional Injuries in the Home.* 2nd ed. http://homesafetycouncil.org/AboutUs/Research/pdfs/sohs_2004_p017.pdf, 2004.

Kohr, Robert L. *Accident Prevention for Hotels, Motels and Restaurants.* New York: Van Nostrand Reinhold, 1991.

Manning and Smith Insurance. "Bathroom Slip and Fall Exposures." *Manning and Smith's Tech Topics*, 2000.

6

The Insurance
Industry and Fraud

Slip-and-fall accidents are preventable and should not be seen as simply a cost of doing business. The property owner who accepts slip-and-fall accidents as a course of business will be forever plagued. The insurance industry, specifically those companies that underwrite the property and casualty polices for business owners, is both the recipient of much of the cost of slips, trips, and falls as well as the very reason many such accidents continue to go unchecked. It's not to say that insurers are the problem, but they are enablers.

One thing the insurance industry is good at is keeping track of losses and their financial costs. However, in the case of fall prevention, they rarely, if ever, take a decisive and proactive approach to prevention. That is, insurers and the insurance brokers simply pay the cost of falls and turn around and pass the cost along to their policyholders, who in turn pass the cost to their customers.

Why is it that the insurance industry has not tried to reduce the growing rate of slips, trips, and falls? Certainly they are aware of the growing frequency and severity of slips and falls. So why don't insurers take a leadership role in rewarding those companies that have developed a proven prevention plan and punish those who haven't? Why not offer a premium discount as an incentive to reward those policyholders who do a good job of reducing losses and in turn increase premiums for those property owners who are habitual safety offenders? As long as the insurance industry continues to pick up the tab for such claims, the problem will only get worse. After all, there is only so much elasticity in the market, and as the nation's recession expands, many business owners will either be forced to drop their coverage or go out of business. Ask any business owner the question, "Do you have a problem with slips and falls?" You will most likely hear, "Sure, that's why we have insurance." Imagine if you asked someone, "Do you have cancer?" I doubt they would answer, "You bet, that's why I have health insurance." Prevention appears to be an alien concept to the nation's insurers, so we continue to accept the ever-mounting cost. It's simply a matter of time before the status quo of increasing insurance premiums will collapse.

In my nearly 20 years in the safety industry, I have collected the top 10 reasons insurers and brokers aren't more aggressive in reducing the number of slips and falls:

1. The market is soft, and we can't do anything at this time.
2. The industry doesn't work that way.
3. That would be up to the broker to do.
4. We don't sell to the client directly; therefore, we can't.
5. It's against our policy to recommend specific products.
6. We offer our policyholders voluntary training classes (note the word "voluntary").
7. None of our competitors are doing it, so why should we?
8. That would have to be done through the (fill in the blank) department.
9. We would if there were a national standard that required us to.
10. We can't.

At the end of the day, the only real answer is, "Why should they?" As long as their policyholders keep paying the cost, why should they care? As the saying goes, follow the money.

COST–BENEFIT RELATIONSHIP

Consider the claim-distribution curve in figure 6.1, which describes slip-and-fall claims losses. Claim costs plotted against frequency show a bell-shaped curve, or normal distribution of claims. With an average claim cost directly in the middle of the bell, the property owner (and underwriter) can expect claim costs within equal ranges of the norm (or one standard deviation). Fewer claims than the anticipated cost range will result in savings to the property owner. Claims greater than the anticipated range are cost overruns or a "crisis."

The traditional method of managing risk seeks to maintain losses within the anticipated cost range while working to minimize future risk to the company. However, this model contains one flaw. It fails to take into account unpaid claims or claims whose monies have been put into reserve. Figure 6.2 describes the "real-world" model. Note the shift in costs. This graph looks much like a water wave in motion that, with every unpaid slip-and-fall claim (or claim in reserve), grows until it becomes a tsunami. If left unchecked, this wave can overpower the organization's ability to effectively manage its bottom line. However, if properly identified and addressed, slip-and-fall claims can be managed and can even become a profit center. This concept is easy to implement, but it is often difficult to convince management that they need to do so.

Slip and Fall Cost Benefit Relationship

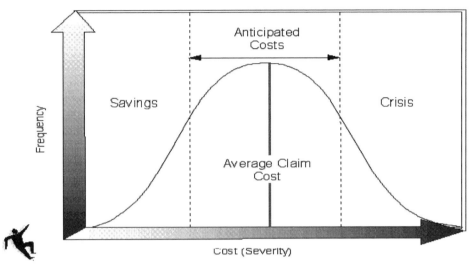

FIGURE 6.1

Slip and Fall Underwriting "Real-World" Model

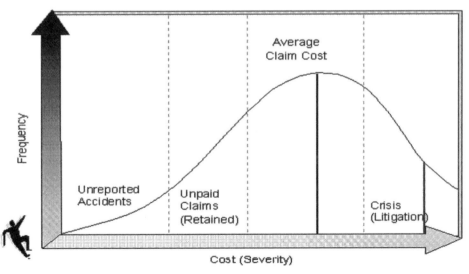

FIGURE 6.2

THE FLOOR SAFETY ASSURANCE MODEL

One method the insurance industry is using to better manage slip-and-fall accident claims is the Floor Safety Assurance Model. This model was first introduced by Steve Spencer of State Farm Insurance and focuses on the following components:

- Product selection (i.e., floor-covering material and maintenance product)
- Using only National Floor Safety Institute (NFSI)-certified products
- Floor testing and frequency (via an independent, third-party auditor)
- Documentation of floor slip-resistance audits (testing)
- Deficiencies and corrective action

A description of the model, written by Steve Spencer, is reproduced in the appendix to this chapter.

LEADING INDICATORS VERSUS LAGGING INDICATORS

For years, safety professionals have relied on different sources of loss data to assist them in identifying risk. Leading indicators such as customer or employee complaints, coupled with lagging indicators such as actual losses or claims, have served an important role in accident prevention. Slips and falls are a bit different in that many slippery surfaces go unreported by customers and employees and become evident only when someone slips and falls. It is not to say that all walkway hazards go unreported; rather, many people simply accept the condition as "normal," like a wet floor in the kitchen of a restaurant, or simply do not take into account the risk that a hazardous walkway presents to other individuals.

FRAUD

The bank robber Willie Sutton once said that the reason he robbed banks was that "that's where the money is." The same can be said for those criminals who target businesses to stage fraudulent slip-and-fall accidents. However, rather than using a gun to pull off their crime, they use staged theatrics, an unsuspecting property owner, and legal loopholes.

No one is sure who the first slip-and-fall con artist was, but what is known is that for at least 100 years, people have been staging bogus slips and falls and filing insurance claims to make cash. Known as "banana peelers," these early con artists perpetrated their fraud by claiming to have slipped and fallen on a banana peel, often on the platform of a commuter train. You see, then, like now, it was common for people to throw food or other trash out of the train cars and onto the passenger decks, only to have boarding passengers slip and fall. One of the earliest known banana peelers was

Mrs. Anna A. Strula (a.k.a. "Banana Anna"), a boardinghouse keeper from Hazlet, New Jersey. Of the 17 known peel-related claims made by Banana Anna, her last fake fall took place on a New York City streetcar in 1912 and wound up sending her to a women's prison in New Jersey.

One of the first recorded fraudulent slip-and-fall claims was conducted by the Freeman family from Chicago, Illinois. The family consisted of a father, mother, and seven children, five of them girls. Having worked in the song-and-dance industry, in 1894 the family began staging accidental falls on banana peels throughout the Illinois Central Railroad. The con was simple: one member of the family would discard a banana peel onto the platform of a train car, and a second would then slip and fall for all the public to see. Unfortunately, like other banana peelers before them, the family got caught. Because of the unusually high number of slip-and-fall claims being made by the family against the railroad company, a suspicious "claims man" sent circulars throughout the country seeking information from other railroads as to the family's slip-and-fall claims history. The replies came pouring in.

During the early 1900s, the typical slip-and-fall claim would settle for a few hundred dollars, sometime as high as $500. The Freeman family of banana peelers might not have ever been caught were it not for their greed. One of the daughters, Fannie Freeman, a veteran of the slip and fall, filed a $2,000 claim for paralysis suffered as the result of a slip and fall on the Chicago Rock Island and Pacific Railway on Christmas Eve 1894. On further investigation, it was soon learned that Fannie was not paralyzed but simply put up to the scam by her mother, who, when confronted by investigators, "grabbed Fannie by the hair, dragged the poor 'paralyzed' thing out of bed, and pounded her vigorously." Needless to say, the family was prosecuted and the ring broken.

Because it was difficult to catch the growing number of banana peelers, the railway claims men began publishing "The Bulletin," a list of would-be con artists. There was no telling just how many other banana peelers were out there, but "The Bulletin" was published through the 1920s and saw a steady drop in banana peel slip-and-fall claims of would-be con artists (Dornstein 1957).

The modern-day association between the banana peel and a slip and fall was first used in a 1920s Vaudeville comedy that depicted a drifter who went about staging fraudulent slip-and-fall accidents with nothing more than a banana peel.

FRAUD FACTS

- Fraud is the intentional misrepresentation of material facts and circumstances to an insurance company to obtain payment that would not otherwise be made.
- Accident frauds come in all shapes and sizes and are not isolated to one type of event or individual. Slips and falls are a common type of fraud, but fraud actually represents less than 3% of all slip, trip, and fall claims.
- Insurance fraud is classified as being either "hard" or "soft." Hard fraud is a deliberate attempt to stage or invent an accident or injury, while soft fraud, sometimes called opportunity fraud, occurs when an individual exaggerates what is otherwise a legitimate claim.

IMPACT OF FRAUD

The insurance industry estimates that at least 10% of all insurance claims are fraudulent, costing an estimated $83 billion (more than the cost of all armed robberies combined). However, only 3% of all slip, trip, and fall insurance claims are considered fraudulent. Surprised? Media coverage of slips, trips, and falls rarely addresses the 97% of legitimate slip-and-fall accidents but rather focuses on the 3% that occur fraudulently. Why? Because it makes for good television. Don't be mistaken: slip and fall con artists cost the insurance industry approximately $2.5 billion each year, but that is far less than the estimated $80 billion to $100 billion spent paying for legitimate falls.

Insurance fraud loss is estimated to be $27.6 billion per year, automobile fraud $12.3 billion, business and commercial fraud $1.8 billion, home-owner fraud $1.8 billion, and life/disability fraud $1.5 billion.

Insurance fraud, the second most costly offense in the white-collar sector, costs the American public approximately $96.2 billion per year in increased premiums alone. A study in 2001 by Conning and Co. estimated that insurance fraud increases the average American household costs by over $5,000 per year when the rise in premiums, goods, and services is taken into consideration. According to the Insurance Information Institute, home-owner fraud, which includes property and casualty claims, totals about $30 billion per year. False claims in the American health care system cost the United States about an $54 billion per year (Coalition Against Insurance Fraud). In Canada, 10% to 15% of claims paid out are fraudulent. The sum of general insurance fraud inflates costs an estimated $1.3 billion per year according to the CCAIF (http://www.insurancefraud.com).

TWELVE SIGNS OF SLIP-AND-FALL FRAUD

So what are the signs that someone has filed a bogus claim? Following are 12 common indicators of possible insurance fraud:

1. The claimant is familiar with the insurance claims and/or legal processes.
2. There are inconsistencies in the claimant's story as to how he or she was injured.
3. The claimant is unemployed and may have a criminal history.
4. The claimant has a filed previous claims for a fall, often in other cities or states.
5. The claimant goes by one or more aliases, has different Social Security numbers or addresses, or refuses to provide an accurate address.
6. The claimant seeks treatment from a physician, hospital, or clinic that is geographically remote from his or her residence.
7. There is difficulty establishing telephone contact with the claimant, especially when the claimant is an employee on medical leave due to injuries.
8. The incident report indicates no evidence of physical injury or no evidence of a hazardous condition.
9. The claimant has worked with the same attorney/doctor combination on a previous claim.
10. There are inconsistencies in the dates of medical treatment as described in physician's notes and billing statements.
11. Medical appointments are scheduled for evening hours or Saturdays when the claimant is allegedly on medical leave due to injuries.
12. The physician discharges the claimant to return to work, but the claimant refuses.

REFERENCE

Dornstein, Ken. *Accidentally, on Purpose: The Making of a Personal Injury Underworld in America.* New York: St. Martin's Press, 1996.

Appendix

RESTAURANT SLIPS-AND-FALLS STUDY: OBJECTIVE AUDITING TECHNIQUES TO CONTROL SLIPS AND FALLS IN RESTAURANTS (PUBLISHED JUNE 2007, CNA INSURANCE RISK CONTROL)

Background

More than 3 million food service employees and over 1 million guests are injured annually as a result of restaurant slips and falls according to the NFSI. The NFSI indicates that the industry spends over $2 billion on such injuries each year and is increasing at a rate of about 10% annually.

According to the National Restaurant Association, slips and falls are the greatest source of general liability (GL) claims within the restaurant industry. CNA's loss results mirror the National Restaurant Association information. Slip-and-falls injuries continue to be the leading source of GL losses incurred by our customers.

Slips and falls, according to the National Safety Council, constitute one of the leading causes of accidental death in the United States with 60% of falls occurring from the same level. With the aging of the baby-boomer generation, the size and scope of this issue is expected to grow significantly. Between 2005 and 2020 it is estimated that the number of seniors will increase from 35 million to 77 million. Statistically, seniors are far more likely to experience a slip-and-fall accident. For those who are injured, the cost of treatment and recovery time are significantly greater than the average. According to the American Academy of Orthopaedic Surgeons, slips and falls are also the leading cause of injury and hospital admission for older adults.

The major causes of slip-and-fall accidents involve five primary sources:

1. Lack of slip resistance on walking surfaces
2. Poor walking surface conditions
3. Poor visibility
4. Lack or poor condition of handrails and guardrails
5. Poor accessibility

This paper will discuss the approach taken by CNA Risk Control to deal with the first two of these causes in a restaurant environment. It will focus on the application of a new technology and a systematic auditing technique to help to objectively identify problem areas and communicate findings and suggestions for improvement from our study to our customer, a national restaurant chain.

Since 2001 our client, a fast-growing national restaurant chain, had identified patron slips and falls as the leading source of GL claims. While the company tried several remedies and experienced some progress in this area, (as measured on a per-store basis), falls continued to serve as their primary "loss leader" from a GL standpoint.

CNA initially began working with the customer on slips and falls issues in April 2004. At that time, a series of floor tests were completed at selected locations using the English XL floor testing device. During this time frame, guest slip-and-fall injuries were confirmed as the primary driving force of the firm's GL losses in terms of both frequency and severity. By December 2004, the firm approached CNA for assistance in developing and implementing a more aggressive slip-and-fall prevention program.

In March 2005 the company had established an internal slip-and-fall prevention program and rolled it out nationwide. Over this same time frame, CNA formed a strategic partnership with the NFSI.

The NFSI was founded in 1997 as a not-for-profit organization dedicated to "aid in the prevention of slip and fall accidents through education, training and research." Headquartered in Dallas, Texas, the firm is the only organization of its kind exclusively focused on slip-and-fall accident prevention.

The NFSI was in the process of developing its B101.0 Walkway Surface Auditing Guideline for the Measurement of Walkway Slip Resistance and submitting it to the American National Standards Institute (ANSI) for standardization. We proposed to use this new standard as part of our study protocol.

In addition, CNA purchased five binary output tribometers (BOTs) to support our commitment to slip-and-fall prevention. Using this technology and the new NFSI walkway auditing guideline, CNA was able to partner with our client to enhance their existing slip-and-fall prevention program. Following site selection, actual testing began at selected sites in July 2006 and concluded in October 2006.

What Is Slip Resistance?

Slip resistance is generally measured by defining the coefficient of friction (COF) between two surfaces—between a shoe and a floor surface, for example. There are two COF measures:

- Static—the force necessary to start one body moving
- Dynamic—the force necessary to keep this same body moving

Here in the United States, static COF is the customary method of measuring slip resistance.

The COF is generally measured between 1.0 for very rough surfaces (i.e., sandpaper) and extremely slippery surfaces at 0.0 (i.e., water on ice).

Food Service General Liability Insurance Claims

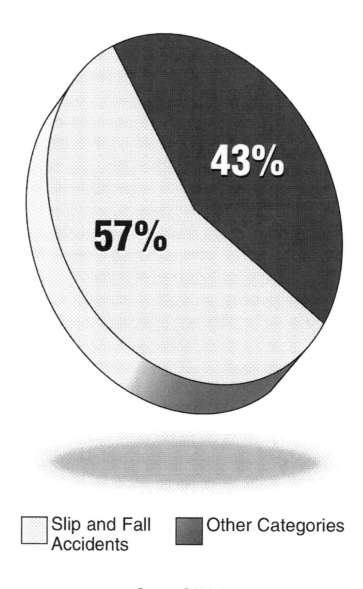

43%

57%

☐ Slip and Fall Accidents   ■ Other Categories

Source: C.N.A. Insurance

FIGURE 6.3

The ANSI A1264.2-2001 "Standard for the Provision of Slip Resistance on Walking and Working Surfaces" suggests a static COF of 0.5 for walking surfaces under dry conditions.

However, the ANSI has developed an additional test method, ANSI/NFSI101 .1-2009. This standard defines a "high-traction" walkway as being one having a measured static COF of .06 for *wet* walking surfaces. According to the institute, floor surfaces maintaining this level of slip resistance when wet have proven to reduce slip-and-fall claims by between 50% and 90%. We chose to use this standard as part of our study, as we felt it more closely replicated real-world situations.

What Factors Influence Slip Resistance?

Any factor that changes the level of friction between two surfaces impacts its slip resistance. When the floor surface and the sole of an individual's shoe are clean and dry, there is generally a high level of friction between the surfaces. In this case, the likelihood of slips and falls is reduced. Over time, as flooring surfaces and shoe soles become covered by foreign materials or become wet, the level of friction is reduced. As this occurs, the likelihood of a slip and fall increases.

We typically think about foreign materials being such things as dirt, grease, and water. However, we also know that some cleaning products used on flooring surfaces can build up a film in the pores of flooring material. This reduces the friction produced by the surface, increasing the likelihood of slips and falls. We call this buildup of materials "polymerization" and know that the longer the buildup continues, the more difficult it is to remove. This becomes extremely important in cases where the floor surface occasionally becomes wet, such as in restaurants.

Frequently in the hospitality industry, we find occasional spills, weather-related hazards, wet and oily surfaces, and changes in the degree of traction as the primary causes of slips and falls.

Our Approach

In preparing for the study, a presentation was made to top management of the restaurant chain. The purpose for the presentation was twofold:

- First, provide education on the slip-and-fall issue and also relay the study's potential benefits to their organization.
- Second, solicit their support and commitment for the project. We also used the session to discuss the equipment and suggest how the sampling could be accomplished.

With management commitment secured, the company notified its managers of the four stores selected for the project and what they could anticipate in terms of the on-site testing. One of the primary objectives of the study was to monitor and document the results of floor-cleaning and maintenance activities, so the decision was made early on to complete readings and measurements during nonbusiness hours. Further, as the primary issue experienced by the firm dealt with customer slip and falls, the decision was made to limit our sampling to only "front-of-the-house" areas of the stores.

We decided to include a series of stores in our study whose layouts and interior finish materials were consistent with what would be included in new stores as the firm expanded across the United States. The sites were also contained within a relatively tight geographic area to allow multiple retesting in an efficient manner.

Owing to our existing relationship, we already understood their market and guest demographics, cleaning and floor maintenance procedures and products, and risk management/slip-and-fall prevention programs. Historical material targeting sites of previous guest slip-and-fall incidents was reviewed and categorized. This information provided a historical perspective to losses and suggested keys to study during the upcoming on-site sampling.

For the purposes of this study, we received on-site assistance from engineers representing Universal Walkway Tester LP, the Dallas-based manufacturer of the equipment used to gather the data (BOT-3000). Their expertise in operational aspects of the equipment and knowledge of the NFSI 101-A floor-auditing guidelines helped us structure the study and interpret the resulting information. Their invaluable assistance with this project was greatly appreciated by the CNA risk control team assigned to the project.

For the purposes of the study, one lead and one backup floor auditor were selected to work with the Universal Walkway Tester LP engineers and the management at the client-selected locations.

What We Did

Our plan included securing two data sets for each location tested. The first samples were obtained after the facility had closed for the evening. The auditor would then return to the site the following morning, after the cleaning crew had completed their work but before the facility had opened for business. "Back-to-back" testing was employed to reduce the possibility of any intervening factors impacting the results of our operational and cleaning protocols.

Our pattern called for testing of two facilities at a time with all sites visited during a two-week period each month. The following month's testing would then be completed the same week four weeks later. In general, testing was completed the same day of the week for each facility in the study, though this was not always the case.

Initial evaluations of the test sites were then completed. During these visits, operations were observed and information was gathered from staff to help determine areas to be addressed in the sampling. Armed with detailed diagrams of each facility, an assessment was carefully completed to identify those locations that would serve as future sampling sites. One important component in the initial testing was to take the time to explain the equipment, purpose, and nature of the testing and potential outcomes to facility staff and management. In each facility, this was their initial contact with the equipment, and it was important that they understood how it worked and what it was used for.

Criteria outlined in NFSI's proposed floor-auditing standard NFSI 101-A, "The Measurement of Walkway Slip Resistance Walkway Surface Auditing Guidelines," served as a resource in location selection. The auditing guidelines subdivide floor surfaces into three groups: (1) normally dry, (2) normally wet, and (3) occasionally contaminated. The occasionally contaminated surface definition best fit the layout and operations present at the facilities selected.

For "occasionally contaminated" categorized floor surfaces, the requirement called for conducting a test using a Neolite® (rubber) sensor on a wet surface. Distilled water was used as the base of each test. Each test consisted of a pair of samples. One taken in an "east-to-west" orientation and the second completed in a "north-to-south" orientation. This allowed us to obtain samples going both "with" and "against" the grain of floor surfaces (where grain was present).

The Testing Process

Based on the layout and arrangement of facilities, between 9 and 13 individual sampling sites were selected for each location. Following our master diagram, subsequent sampling was to be completed at these specific sites. Areas tested included dining/seating, bar, beverage stations, serving routes, restrooms, hostess, and entrance/exit points. Special attention was paid to high-traffic area locations where different flooring materials met. These transition areas frequently were a source of slip-and-fall incidents.

Ultimately, a pattern was established of the auditor arriving at the store 15 to 30 minutes prior to closing. This was done to prepare and validate the equipment before each day's testing as well as to observe operations and determine if any additional information could be obtained regarding the firm's customer slip-and-fall injury trends. Once the facility was closed and free of guests, sampling began. Generally each set of samples would take 5 to 10 minutes to complete. During this time, sample media needed to be prepared, testing surfaces prepared with distilled water, samples run, date recorded, and sample sites cleaned up of water.

Subsequent testing the following morning generally went more quickly as testing commenced following validation of the equipment on arrival at the site.

What We Used

We used a BOT Model 3000, manufactured by Universal Walkway Testing LP. The BOT-3000 is a self-propelled BOT, based on a modified drag-sled principle. The unit can perform both wet- and dry-surface testing. The digital instrument can record, print, and output data that can be cataloged and analyzed. One key feature of the unit is that it can be field calibrated, and the results are user independent.

What We Learned

Following manufacturers' exact directions when applying floor-cleaning compounds is crucial to the success of a floor maintenance program. Proper training and outfitting of applicators must be monitored. Targeting cleaning and floor maintenance activities to those areas known for producing low slip resistance make a slip-and-fall prevention program more efficient.

Aggressive use of floor mats can greatly reduce the tracking of materials such as grease around the facility. Truly effective mat programs feature frequent change-out schedules so that the mats themselves don't become a medium for moving contaminants throughout the store.

Understanding the individual store's loss history was very important in the planning of the study. To this end, detailed incident reports identifying flooring material, location, time of day, nature of the incident, and information on the claimant are important factors that can help determine where and when to sample. The information will also help train staff on situations and conditions likely to result in an incident.

Sampling in consistent locations, month after month, both before and after cleaning, provided us good information on the success and challenges faced by each facility's floor care and maintenance program. Having specific site diagrams outlining sampling locations aided in the consistency of collecting information.

Following a consistent pre- and postsurvey process helped ensure the proper operation of the equipment and consistency of the results achieved.

Even though the flooring surfaces, facility layouts, operations, and cleaning products used were consistent over the locations involved in the study, there were considerable differences in the slip resistance readings between locations. The common difference that each facility shared was that floor maintenance and cleaning was performed by an outside contractor. Controlling for all other factors, contractor application emerged as a critical variable in the process.

Following cleaning, each flooring surface exhibited a significant improvement in its individual slip resistance. The actual degree of improvement differed with each facility and sampling location. This was noted to be especially true in heavily contaminated areas, such as entrances to the kitchen, food preparation areas, and beverage stations. Also, the improvement was generally consistent in sample areas, measured on a month-to-month basis.

The tracking of materials such as grease, oil, and water from the kitchen to the serving and seating areas emerged as the primary controllable source of improving overall slip resistance. Our slip-resistance readings consistently improved the farther we moved away from the entrance to the kitchen and/or serving areas.

Floor mats, especially used at the entrance to the kitchen and serving areas, effectively reduced the movement of materials such a grease and water from other portions of the facility. To maintain their effectiveness, the mats needed to be changed at regular intervals before becoming saturated. We observed that saturated mats can make the situation worse by adding to the problem.

Similarly, areas with permanently installed mats and carpet runners need to undergo regular maintenance and cleaning to remove the buildup of materials that could otherwise be tracked throughout a facility.

Advanced preparation was essential to a smooth-running sampling session. Validation and preparation of the equipment helped to reduce required on-site time. Being properly outfitted with supplies reduced delays on site.

Employing separate color-coded mops and buckets for "front-of-the-house" and "back-of-the-house" areas helped reduce cross contamination. Using mops that typically are used in the kitchen and preparation areas in the customer seating areas are a frequent source of the spread of slip-resistance-lowering materials.

Recommendations

- Select slip-resistant materials when you build, expand, or remodel your facilities. Installation of materials with proven high-traction characteristics is one of the best ways to avoid slip-and-fall issues later on.
- Know what the "out-of-the-box" slip resistance is on the floor materials you have in your facility. These numbers provide a baseline when considering changes to your cleaning and floor maintenance practices.
- Select floor-cleaning and maintenance products with proven slip-resistance characteristics that are compatible with the particular flooring surfaces in your facility. A good place to start is materials certified by the NFSI (http://www.nfsi.org).
- Be alert for workers substituting cleaning materials or supplies. Ensure sufficient supplies of materials.
- Apply your floor-cleaning and maintenance products in accordance with the manufacturer's recommendations. Verify with the cleaning personnel that they are familiar with and are using the correct application procedures. Observe application procedures if there is a change in personnel or contractor used for this service.

- Remove any unauthorized or incompatible cleaning products and educate your staff to the potentially dangerous consequences that using the wrong products can have on the slip resistance of your flooring surfaces.
- Separate cleaning materials and equipment between the "front of the house" and "back of the house" to reduce the likelihood of transporting a problem from one area to another. Color coding materials can provide instant recognition of personnel using the wrong equipment in the wrong area of the facility.
- Ensure that permanently installed features like carpet runners and mats are included in your maintenance and housekeeping program. These materials need to be regularly inspected for the buildup of contaminants and deterioration that could lead to the creation of fall hazards. Keep in mind that while the mats reduce the likelihood of producing slips, improperly maintained mats can create trip hazards.
- Limit the difference in heights between flooring surfaces and mats to no more than one-quarter to one-half inch. Make sure to evaluate the condition of these changes in height, as they can deteriorate and create trip hazards.
- Regularly review all the slip-and-fall incident reports associated with your facility and understand the critical factors associated with them. Look for trends in location, time of day, and so on and focus staff training on your cleaning procedures toward these factors. Train your workers how to properly respond to slip-and-fall incidents.
- Ensure that the staff is well trained in your spill prevention and response programs. They need to know where the materials are located and how to use them in the event of an emergency. It's also important that the staff understand the importance of reporting incidents and conditions that could result in incidents, even if none has actually occurred. These will be your first indication of a potential issue that should be addressed.
- One of the surest ways to "short-circuit" the transmission of grease, water, and other materials from the "back of the house" to the "front of the house" is to implement a good mat program. Ensure the mats are checked regularly for wear and the buildup of contaminants. A poorly managed and maintained mat program can significantly increase the likelihood of reducing the slip resistance of your flooring surfaces.
- A floor-auditing program can help you identify trends within your facility that can result in reduced slip resistance to your flooring surfaces. To be effective, the testing should be completed in a consistent manner and include more than a single set of measurements.
- Maintaining open communication between the staff, cleaning personnel, and the floor auditor is crucial to the identification of trends and the elimination of factors that could reduce the slip resistance on your floor surfaces.

7

The Legal Industry

Who's afraid of the big bad wolf? When the wolf is an attorney, just about everybody. The subject of slips and falls is a very important subject to property owners, flooring manufacturers, and building service contractors. Why? Because none of them want to be on the receiving end of a lawsuit.

However, rather than facing the underlying causes of most slip-and-fall accidents (i.e., a hazardous floor), many floor-covering and floor care manufacturers simply pass the slip-and-fall problem along to their customers (i.e., business owners), who wind up paying higher insurance premiums and in the end pass the higher cost of doing business along to their customers. If I only had a dime for every time I heard people tell me, "Sure we have a few slip-and-fall lawsuits, that's why we have insurance"—the thought being that somehow the insurance industry can pay out tens of thousands of dollars in legal fees and not pass the cost along to them, the policyholder. This merry-go-round of passing the cost along to the next guy has inadvertently served to discourage property owners, product manufacturers, and building service contractors from taking a proactive approach to slip-and-fall prevention, and it ends up contributing to the overall growing epidemic of slip-and-fall accidents.

According to the Jury Verdict Research, Insurance Information Institute, plaintiffs win 51% of premises liability claims, and million-dollar-plus jury verdicts increased 13% from 1990 to 2000.

Overall Industry Specific Data

Majority of Premises Liability Verdicts are greater than $100,000

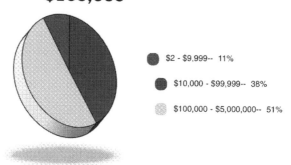

- $2 - $9,999-- 11%
- $10,000 - $99,999-- 38%
- $100,000 - $5,000,000-- 51%

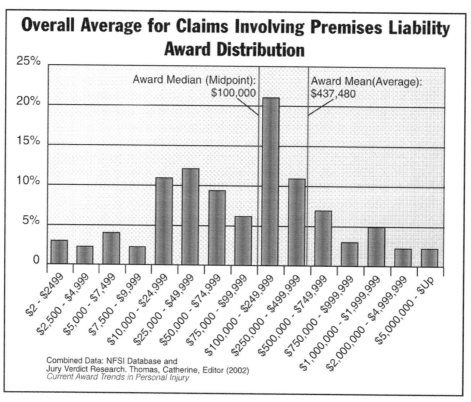

FIGURE 7.1

Majority of Hospitality Premises Liability Verdicts are Greater than $75,000

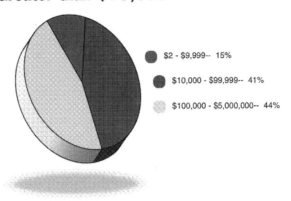

- $2 - $9,999-- 15%
- $10,000 - $99,999-- 41%
- $100,000 - $5,000,000-- 44%

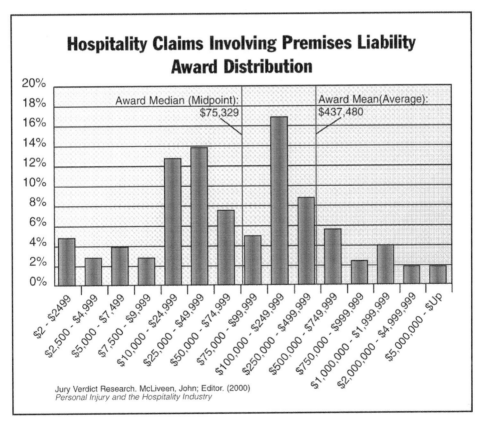

FIGURE 7.2

Majority of Industrial Premises Liability Verdicts are Greater than $280,000

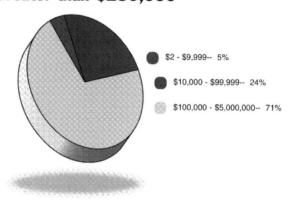

● $2 - $9,999-- 5%

● $10,000 - $99,999-- 24%

○ $100,000 - $5,000,000-- 71%

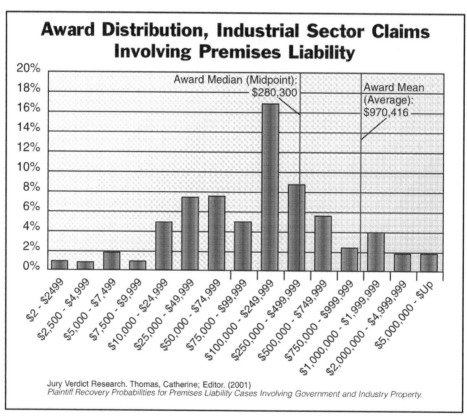

Award Distribution, Industrial Sector Claims Involving Premises Liability

Award Median (Midpoint): $280,300

Award Mean (Average): $970,416

Jury Verdict Research. Thomas, Catherine; Editor. (2001)
Plaintiff Recovery Probabilities for Premises Liability Cases Involving Government and Industry Property.

FIGURE 7.3

Majority of Government / Schools Property Premises Liability Verdicts are Greater than $119,000

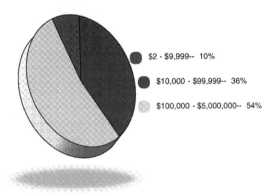

● $2 - $9,999-- 10%

● $10,000 - $99,999-- 36%

○ $100,000 - $5,000,000-- 54%

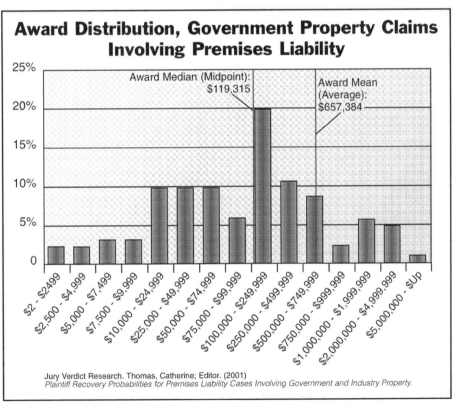

Award Distribution, Government Property Claims Involving Premises Liability

Award Median (Midpoint): $119,315

Award Mean (Average): $657,384

25%
20%
15%
10%
5%
0

$2 - $2499
$2,500 - $4,999
$5,000 - $7,499
$7,500 - $9,999
$10,000 - $24,999
$25,000 - $49,999
$50,000 - $74,999
$75,000 - $99,999
$100,000 - $249,999
$250,000 - $499,999
$500,000 - $749,999
$750,000 - $999,999
$1,000,000 - $1,999,999
$2,000,000 - $4,999,999
$5,000,000 - $Up

Jury Verdict Research. Thomas, Catherine; Editor. (2001)
Plaintiff Recovery Probabilities for Premises Liability Cases Involving Government and Industry Property.

FIGURE 7.4

Majority of Food Service Liability Verdicts are Greater than $85,000

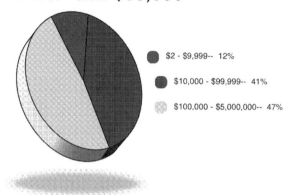

- $2 - $9,999-- 12%
- $10,000 - $99,999-- 41%
- $100,000 - $5,000,000-- 47%

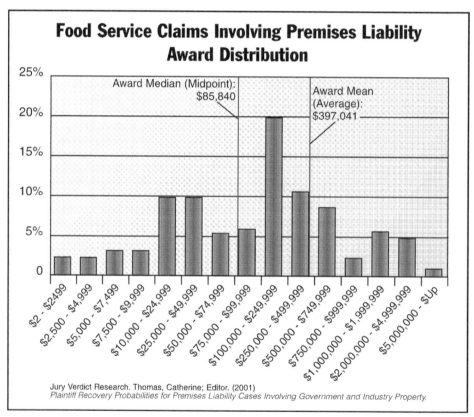

FIGURE 7.5

Majority of Building Service Premises Liability Verdicts are Greater than $78,000

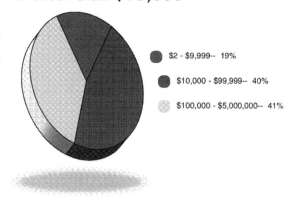

$2 - $9,999-- 19%

$10,000 - $99,999-- 40%

$100,000 - $5,000,000-- 41%

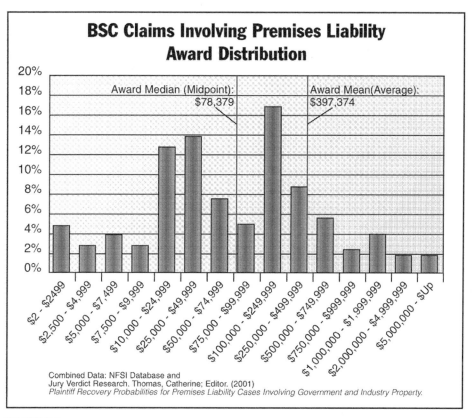

BSC Claims Involving Premises Liability
Award Distribution

Award Median (Midpoint): $78,379

Award Mean(Average): $397,374

Combined Data: NFSI Database and
Jury Verdict Research. Thomas, Catherine; Editor. (2001)
Plaintiff Recovery Probabilities for Premises Liability Cases Involving Government and Industry Property.

FIGURE 7.6

Majority of Retail Store Premises Liability Verdicts are Greater than $50,000

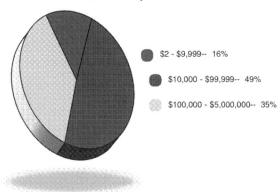

● $2 - $9,999-- 16%

● $10,000 - $99,999-- 49%

○ $100,000 - $5,000,000-- 35%

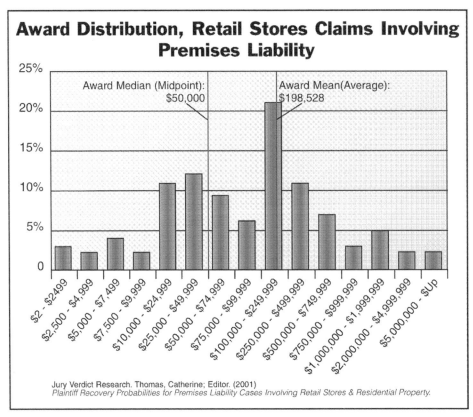

Award Distribution, Retail Stores Claims Involving Premises Liability

Award Median (Midpoint): $50,000

Award Mean(Average): $198,528

$2 - $2499
$2,500 - $4,999
$5,000 - $7,499
$7,500 - $9,999
$10,000 - $24,999
$25,000 - $49,999
$50,000 - $74,999
$75,000 - $99,999
$100,000 - $249,999
$250,000 - $499,999
$500,000 - $749,999
$750,000 - $999,999
$1,000,000 - $1,999,999
$2,000,000 - $4,999,999
$5,000,000 - $Up

Jury Verdict Research. Thomas, Catherine; Editor. (2001)
Plaintiff Recovery Probabilities for Premises Liability Cases Involving Retail Stores & Residential Property.

FIGURE 7.7

QUICK FACTS

- According to the American Bar Association, lawyers specializing in personal injury cases represent the fastest-growing segment of the legal profession and currently represent more than 30% of all practicing attorneys.
- Jury accident awards of $1 million have increased from three per year in 1960 to more than 600 in 1987, and the trend continues.

EXPERT WITNESSES AND PROFESSIONAL EXPERTS

About 10 years ago, a cottage industry arose around the expert witness. What exactly makes one an expert witness? Are you an expert based on your level of education, or is your work experience in the field of safety more important? Does it matter how many books and articles you have published? The answer is all of the above. When qualifying an expert witness in the field of slips, trips, and falls, it is important to have the right combination of training, experience, and professionalism, not to mention real-world experience in preventing such accidents. For example, just because someone has a degree in engineering does not make him or her an expert in all areas of engineering, and just because an individual has the initials CSP (certified safety professional) after his or her name does not qualify him or her as an expert in all areas of safety, nor does it speak to his or her level of professionalism.

In his book titled *Become a Recognized Authority in Your Field in 60 Days or Less*, Robert W. Bly (2002) discusses identifying an expert as a guru and asks, "Are gurus B.S. artists?" Who qualifies for guruship? What are the benefits of becoming a guru? In the area of expert witness, there are few gurus; rather, there is a long line of retired engineers, burned out MDs and downsized middle managers. What all these groups have in common is their desire to be their own bosses, earn big money, and work in the exciting and growing legal profession.

For the past 20 years that I have worked in the field of slip-and-fall accident prevention (first as a product manufacturer and then as founder of the National Floor Safety Institute [NFSI], therefore qualifying me as an expert in the field of slip-and-fall accident prevention), I have met many highly qualified and honest expert witnesses and even more self-promoting charlatans. No accredited university offers a degree in slips and falls, and little formal training on the subject is even available. For some so-called experts, they hold out their degrees in engineering, often PhDs, which is impressive; however, most if not all of their formal training was in areas of engineering totally unrelated to walkway safety. Just because you have a PhD in aerospace, mechanical, or other engineering disciplines does not make you an expert in slips, trips, and falls.

Over the past 12 years, I have been retained on over 400 slip, trip, and fall–related lawsuits and have turned down nearly the same number of cases. My workload, like many in my field, often comes by way of referrals from attorneys who have hired me in the past and recommend me to other attorneys. My caseload has been almost equally split between plaintiffs and defendants and I have worked on a wide range of cases across the United States and Canada. I am not a plaintiff advocate or a shill for the defense but rather an advocate for pedestrian safety. For the past 15 years, I have volunteered thousands of hours of my professional time to serving on not-for-profit safety boards and standards-developing committees and wholeheartedly believe that my contribution to society has led to making the world a safer place. It's not a matter of whether people will become injured as a result of a fall but rather a matter of who, if anyone, is responsible for the injuries. The role of the expert witness is to assist the trier of fact (jury) in understanding what may be a complex set of circumstances and make what is complex simple. If one party was in fact negligent, that must be supported with more than opinions. Industry standards, trade practices, and published safety guidelines all serve important roles in determining liability or negligence. An individual who relies on little more than his or her pedigree and private collection of "authoritative literature" is often of little help. These "professional experts" are often of little assistance to the trier of fact and often bring more confusion than clarity to the legal process.

Attorneys Need Experts, and Experts Need Attorneys

The symbiotic relationship between experts and attorneys has never been as important than in cases involving falls. Just how and why people fall is not an exact science but rather a pseudoscience. Experts who can apply scientific techniques, applicable test methods, and professional work experience can be valuable to both plaintiffs and defendants.

Courtroom Testimony

Q: Before you performed the autopsy, did you check for a pulse?

A: No.

Q: Did you check for blood pressure?

A: No.

Q: Did you check for breathing?

A: No.

Q: So, then, it is possible that the patient was alive when you began the autopsy?

A: No.

Q: How can you be so sure, Doctor?

A: Because his brain was sitting on my desk in a jar.

Q: But could the patient have still been alive, nevertheless?

A: Yes, it is possible that he could have been alive and practicing law.

Here is an example of expert testimony misstating the application of industry standards:

Q: Okay. So ANSI does not define what a coefficient of friction for a wet or contaminated floor should have been, does it?

A: ANSI doesn't, that's correct.

Q: Okay. Is it a promulgated standard upon which you relied that it does specify a coefficient of friction for a wet or contaminated floor?

A: In terms of a standard, I think you have to understand what standards are. Standards kind of skim some things that are in the authoritative literature and make them universal.

Q: Okay. Standards—life goes on beyond standards, so they haven't gotten to the point that they will put—you know, put a level on wet testing. That doesn't mean there isn't criteria for wet testing. That doesn't mean the literature doesn't contain concepts and material that you can use for wet testing.

A: Certainly, ASTM has two instruments that they approve for wet testing. The cutoff coefficient of friction for wet testing is the same as for dry testing because it's not an instrument criteria. It's not a—even a physics criteria. It's not a mathematical criteria. It's a human criteria. In other words, the .5 is arrived at from epidemiology. You have to go to back to the source. The .05 has been designated as .5 and authenticated in the authoritative literature because above .5, historically, that's where people fall—that's where people don't fall. Below .5, that's where people fall. That is the source of the .5. It's not some magical mathematical calculation, so—so you have to look at the source.

Q: Okay.

A: Now, the—the wet testing, most of the instruments that have done wet testing in the past are—have been shown to be—have some error in them. So wet testing—accurate wet testing—is recent over the last 10 years maybe. So the standards have not yet put their stamp of approval on any level for wet testing. That doesn't mean it doesn't exist. It's the same as for dry because the—the criteria is does your foot slip? The instruments that we use today—the authoritative approved instruments—test whether your foot slips whether it's wet or dry, and the criteria is the same. There's no reason to believe it's any different than .5. So merely because the standards haven't adopted something—merely they're waiting to adopt something, doesn't mean it doesn't exist.

I might add in terms of analogy that the American Red Cross did not approve CPR for 15 years because they were afraid it was too complex. But in this aspect I had the opportunity to speak with a physician who created this technique; that he was almost in tears that how many—so many lives could be saved if the Red Cross would just teach it. He said you can teach it in five minutes. You can teach it in a 30-second TV commercial and it saves lives. But they wouldn't accept it for 15 years.

So you've got to understand that these standards organizations don't want to make something universal even though it's in the authoritative literature. Even though it can be backed by physics and experience and epidemiology, they wait until the last—they're the last ones to put it into writing. So because it isn't here, doesn't mean it doesn't exist.

Here's another example of expert testimony; this time, the expert misapplies human factors issues with industry standards:

Q: Okay. And I think we know that Ms. X was moving at the time she claims she slipped, correct?

A: Yes.

Q: And as you walk, your weight shifts from one foot to the other just in general, correct?

A: In general, yes.

Q: At some point during your stride, your weight could be on one foot, the other, or equally distributed or some portion of distribution between the two, correct?

A: Correct.

Q: Do you know anything about at what point in her stride Ms. X was when she slipped?

A: The vast majority of slips occur at heel strike.

Q: Okay.

A: I have no reason to believe that it was any different

Q: But she didn't tell you or provide you with any information of that fact?

A: I don't think anybody knows . . .

Q: Right.

A: . . . or could find out.

Q: And that's my point. I don't think—and I think that's correct. Nobody knows exactly what she did when she fell.

A: The other—the other—the other occurs at what they call push off. This paragraph here, for clarification, is talking about someone might on a floor that's, let's say, is .65—it would test to be acceptable—something happens and they—they fall. They may think they slip, but they may not. It may be some other reason. I think that's what they're—they're talking about here.

Q: Well, and that's kind of my point here. Can you say—can you tell this jury that Ms. X would not have slipped if the coefficient of friction of this floor had been higher?

A: With a high degree of scientific certainty, yes.

Q: How can you do that?

A: Because what they're talking about here is one in millions of steps. And what you're saying is, well, even in the presence of an excessively slippery floor, she just went down for a random reason. I mean you argue that; I'm not.

Q: Okay. Well, I mean I'm just saying how do you rule out . . .

A: You don't.

Q: . . . how do you rule out that she just had a misstep?

A: You don't. You don't rule out at a hundred percent. But with a high degree of engineering and scientific certainty, that didn't happen.

Here is a good example of hazard control as described by an expert witness:

The accident process can be visualized by the game of dominos that we may have all played where you put dominos—stand dominos next to each other, and then you hit one and they all fall down including the last one. Well, the last one is the injury. The one preceding it may have been some circumstances that the injured was involved with. But then if you go back in time, here's where you look at the design of the environment where the accident took place. And based on the two principles, first you try to eliminate hazards. Second, you try to safeguard hazards. Well, and third you develop procedures.

But if we had these dominos here and we said, okay, this domino can be removed. Okay. We can put in an inherently slip-resistant floor even if it's wet, and we can just remove the hazardous flooring in this matter. So that's the concept of removing a hazard.

Now, let's say we had—let's take not a workplace, but let's take a bathroom in a hotel. You may have a slippery floor, and you've got water there. So what—on the second rule of hazard control, you add something to the system. Well, what can you add to the system? Well, you can add a handrail. You can't use that here in the workplace; there's no place for it. But in a—in a hotel bathroom you can put a handrail. That's adding something to the system, and that's just like gluing one of those dominos down. So if you can remove a domino or glue it down, once they start to fall, the accident process will stop either where the domino had been removed or where the domino was glued down, and it's the same thing in the real world. The accident process will stop if you can remove the hazard or if you can safeguard it. So that's how you use the accident process, not only to investigate accidents but to prevent them in the first place.

LEGAL TERMS

- *Negligence.* The failure to use ordinary care, that is, failing to do that which a person of ordinary prudence would have done under the same or similar circumstances or doing that which a person of ordinary prudence would not have done under the same or similar circumstances.
- *Ordinary Care.* The degree of care that would be used by a person of ordinary prudence under the same or similar circumstances.
- *Proximate Cause.* That cause which in a natural and continuous sequence produces an event and without which cause such event would not have occurred. In order to be a proximate cause, the act or omission complained of must be such that a person using ordinary care would have foreseen that the event or some similar event might reasonably result therefrom. There may be more than one proximate cause of an event, but if an act or omission of any person not a party to the suit was the "sole proximate cause" of an occurrence, then no act or omission of any other person could have been a proximate cause.

DUTY OF CARE/STANDARD OF CARE

Wet-Floor Signs/Hazard Identification

How often have we all seen wet-floor signs posted on floors that are not wet? Often. In a 1998 study by the NFSI, it was found that 65% of all wet-floor signs are posted on floors that are not wet, and this, over time, negates the real meaning and importance of hazard identification, thus leading to viewer apathy. We see wet-floor signs so often that we have come to ignore them. The floor sign manufacturers have recognized such and have responded by making their products larger and in different colors, thinking that if we see a hot-pink wet-floor sign, somehow we will pay attention, which we do, that is, until we see the hot-pink sign out at times the floor is not wet.

In fact, it is the industry standard to post wet-floor signs as a first line of defense in a slip-and-fall lawsuit. It is common for building service contractors and janitors to put wet-floor signs out all day regardless of whether the floor is wet or not. A good example is that of airport restrooms. Have you ever noticed that wet-floor signs are *always* posted outside the entrance to an airport restroom? Why? The reason is that a floor in the restroom is likely to become wet; therefore, it is important to warn of such a hazard. The problem is that wet-floor signs are to be used temporarily and are not meant to be permanent fixtures. The purpose of a wet-floor sign is to provide a warning to pedestrians that an impending hazard (i.e., a wet floor) is present that the property owner will rectify (i.e., dry the floor). Once the floor has been dried, the sign should be removed, although in the real world this is not how wet-floor signs are used. Because of the fear of litigation, many property owners have opted to post wet-floor signs as the first and only defense in accident prevention, foolishly believing that as long as a wet-floor sign is posted, they cannot be seen as negligent.

EXAMPLE OF A SLIP-AND-FALL LEGAL COMPLAINT

While staying in the defendants' motel, the plaintiff was carefully exiting the shower when she slipped and fell violently to the floor. As a result of her fall, the plaintiff received serious injuries to various parts of her body.

1. Plaintiff was exercising ordinary care at all times material to this action.
2. At all times material to this action, the plaintiff was an invitee of the defendants, and therefore the defendants owed the plaintiff a duty of care.
3. Defendants, individually and/or through their agents, knew or should have known of the existence of the hazardous condition.
4. The defendants' negligence was the direct and proximate cause of the plaintiff's injuries.
5. The defendants were negligent in the following respects:

    (a) In failing to exercise ordinary care to maintain the premises in a safe condition
    (b) In failing to adequately inspect the premises to discover the hazardous condition
    (c) In failing to correct the hazardous condition
    (d) In failing to warn the plaintiff of the existence of the hazardous condition

As a direct and proximate result of the defendants' negligence, the plaintiff has suffered severe bodily injuries requiring medical treatment. The plaintiff did then suffer, does now suffer, and, in all reasonable probability, will continue to suffer in the future extreme physical pain.

CAP ON LAWSUIT DAMAGES UPHELD—RIGHT TO A JURY TRIAL DOES NOT RULE OUT $350,000 LIMIT

Ohioans who sue over injuries or defective products can, with rare exceptions, forget about multi-million-dollar payouts.

The Ohio Supreme Court ruled yesterday that a Republican-led effort to clamp down on big monetary awards by juries does not violate the constitutional guarantee of a jury trial. Business groups hailed the court's 5–2 ruling as a victory for the Ohio economy. They said that runaway awards over such things as defective medications, unsafe cars and slip-and-fall accidents were chasing businesses out of Ohio before the state legislature voted to cap damages in 2005.

But trial lawyers said the verdict will make it easier for companies to market faulty products and escape liability. The case began after a Cincinnati-area woman, Melisa Arbino, experienced life-threatening blood clots from using a birth-control patch marketed by the pharmaceutical giant Johnson & Johnson.

Arbino sued the New Jersey-based company, and before her case would go to trial, her attorneys challenged Ohio's limits on noneconomic damages from product lawsuits. The law caps awards at $250,000 or three times the amount of economic damages, whichever is greater, up to a total limit of $350,000 per plaintiff or $500,000 per occurrence. The limits don't apply in cases where someone suffers permanent disability or loss of a limb or organ system.

Arbino's attorney, Jane G. Abaray, said the limits help unscrupulous businesses but do nothing for the state's economy. "They don't have to be a business in Ohio to benefit from this decision," Abaray said. "All they have to do is hurt people in Ohio."

The Supreme Court decision upheld the legislature's 2005 lawsuit reforms but did not speak to whether Arbino and roughly 2,000 other women suing over Johnson & Johnson's Ortho Evra patch have a valid claim. That issue is pending in the U.S. District Court for the Northern District of Ohio in Toledo. Johnson & Johnson, which still sells the birth-control patch, said it's safe when used as prescribed.

Business groups and the chief sponsor of the 2005 law, state Sen. Steve Stivers, R-Columbus, said the Supreme Court ruling is a boon to Ohio businesses and ultimately should lead to lower insurance rates for consumers. "It recognized that any one of us could be a victim at any time and made sure it covered all of the economic losses and as much as one-half of a million dollars for noneconomic damages," Stivers said. "The jury still has the freedom to decide what they think is fair for them, and then the limits are imposed after the damages are awarded."

Abaray scoffed at that argument, saying a multi-million-dollar verdict is meaningless if a company doesn't have to pay it.

Ty Pine, vice chairman of the pro-business Ohio Alliance for Civil Justice, said the Supreme Court decision should assure businesses that they won't be hammered with enormous verdicts. "For the first time, Ohio businesses and citizens can enjoy the finality of this debate," Pine said. "As we begin in Ohio to move toward more reasonableness in terms of awards in these types of cases, we will hopefully be able to enjoy lower insurance rates."

Trial lawyers have panned the notion of a litigation crises, noting that the Supreme Court's own statistics show that the number of cases alleging defective products and malpractice by doctors and other professionals has declined in Ohio during the past eight years.

The Supreme Court's decision affects cases involving allegedly defective products and medications, as well as wrongful-death, injury and employment-discrimination claims. It does not apply to medical-malpractice lawsuits, although the General Assembly also has capped damages in those cases.

The decision was written by Chief Justice Thomas J. Moyer and joined by Justices Evelyn Lundberg Stratton, Judith Ann Lanzinger and Maureen O'Connor. Justice Robert R. Cupp entered a separate opinion agreeing with the decision. Justices Paul E. Pfeifer and Terrence O'Donnell entered separate dissents. All seven justices are Republicans, although Pfeifer often sides with more liberal interests.

Moyer wrote that many states have reached different conclusions about whether to restrict awards in defective-product and malpractice cases, but that the Supreme Court had no basis for overturning the Ohio law. "This court is not the forum to second-guess such legislative choice; we must simply determine whether they comply with the Constitution," Moyer wrote.

Pfiefer wrote that the court's ruling would allow the legislature to set any limit on damages, which would make a mockery of the jury system's ability to determine awards on a case-by-case basis. "Under this court's reasoning, there is nothing in the Ohio Constitution to restrain the General Assembly from limiting noneconomic damages to $1," Pfeifer wrote. "In essence, the power to cap noneconomic damages is the power to eliminate them" (James Nash, *Columbus Dispatch*).

EMERGING LEGISLATION

The state of Florida recently passed House Bill No. 689, effective July 1, 2010, which nullifies the holding in *Owens v. Publix Supermarkets, Inc.* (802 So. 2d 315, Fla. 2001) and repeals 768.0710 F.S. The new law, 768.0755, requires that a plaintiff prove that the defendant (usually a business establishment or property owner) had actual or constructive knowledge of the dangerous condition that caused injury and should have taken action to remedy it. The new law eliminates the burden on the defendant to produce evidence that they exercised reasonable care under the circumstances.

So how exactly can a victim of a slip-and-fall prove that the property owner knew of the hazardous condition prior to their fall? Wihtout surveillance video, it's very difficult. The nature of this new law defies logic. No other industry has a state law designed to protect them from lawsuits that arise from their unsafe business practices. Imagine if the state of Michigan established a law that required victims whose braking system failed to prove that the car manufacturer knew that the brakes would fail prior to selling them the car. The point is not that the manufacturer knew of the hazard; it's that they *should* have known.

The attempt by the state of Florida to reduce frivolous lawsuits is a lot like chemotherapy treatment for cancer victims: it intends to destroy the bad cancer cells but destroys a lot of good cells in the process. In America, safety is large ensured by litigation. The threat of a lawsuit is an incentive to manufacturers and property owners to ensure that their products and properties are safe. Removing the threat of litigation will only serve to make things a little less safe—and the consumer will ultimately pay the price.

The appendix to this chapter reproduces eHow.com's advice on How to File a Slip-and-Fall Lawsuit.

REFERENCE

Bly, Robert W. *Become a Recognized Authority in Your Field in 60 Days or Less.* Indianapolis: Alpha Books, 2002.

Appendix

HOW TO FILE A SLIP-AND-FALL LAWSUIT
(FROM EHOW.COM, WRITTEN BY JOHN D. MCMAHAN AND G. BRENT BURKS)

Thousands of slip-and-fall accidents occur every year. These personal injuries can happen in public places or an individual's home. The victim's injuries may be minimal, or they may be very serious, resulting in broken bones, paralysis, or even death. The injuries may prevent you from returning to work. Many of these cases result in a lawsuit. Read on to learn how to proceed if you think you have a lawsuit.

Instructions

- Step 1. File your lawsuit before the statute of limitations expires. The statute of limitations (SOL) varies from state to state. It's typically one or two years.
- Step 2. Prepare for the lawsuit by preparing all pertinent information, such as medical care and treatment expenses. Submit photos immediately to justify the legitimacy of your claim and value.
- Step 3. Acquire all the contact information and statements from witnesses immediately. Witnesses can move out of the area, and over time they may forget some of the important details of the accident, so it's best to get this information right away.
- Step 4. Determine if the property owner is liable for your slip-and-fall accident.
- Step 5. Show that the property owner caused the unsafe environment or condition, usually by spilling something and not cleaning it up or by leaving something on the floor in a heavily traveled area.
- Step 6. Prove that the property owner was aware of the situation but neglected to block off the unsafe area or post a warning sign.
- Step 7. Focus on the fact that the property owner should have known about the danger because any reasonable person would have recognized this to be a problem and taken the proper actions to prevent any injuries.

Do not attempt to file your slip-and-fall lawsuit after the SOL deadline.

PERFECTING SLIP-AND-FALL LITIGATION

By selecting a strong case, hiring the right experts, and allocating fault appropriately, you can obtain justice for injured plaintiffs.

Trial: December 1, 2001
Attorney: John D. McMahan

Many plaintiff lawyers refuse to take slip-and-fall cases. They argue that these claims are too time consuming and costly to prepare. But with the right case-screening criteria and premises liability experts, you can defeat many of the legal defenses that often prevent these cases from going before a jury, making it well worth your time and effort to handle them.

Jurors have been receptive to slip-and-fall cases where the injuries are serious and the liability has been clearly established by reliable, credible experts. Appropriate screening criteria will help you identify clients who have extensive, permanent injuries—not just connective tissue injuries. (It is often difficult to convince jurors that a plaintiff with a connective-tissue injury is severely hurt.)

Well-selected cases involve issues of corporate responsibility, which jurors can readily understand, having experienced many business-premises hazards as consumers. With the right facts and presentation, jurors may embrace the plaintiff's corporate-responsibility arguments and award appropriate damages for pain and suffering, loss of enjoyment of life, and permanent impairment.

Most slip-and-fall cases involve one of the following safety hazards:

- *Defect on Business Premises.* Examples include merchandise that has been stacked too high or left on the floor by employees, flooring defects, and improperly attached safety bars in showers.
- *Dangerous Condition or Process.* Examples include food spilled on the floor around a self-serve salad bar, liquid leaking from a machine, misplaced floor mats, and an unfenced pool.
- *Failure to Properly Maintain an Otherwise Safe Property, Leading to Dangerous Conditions.* Examples include loose handrails and wooden stairs that need to be replaced.

Ask the following questions as you consider whether to accept a slip-and-fall case:

1. Did the defendant cause or create the defect or dangerous condition? This may sound simple, but clients often do not know what made them fall. The plaintiff must know or must have a witness who can identify the defect or danger. If the potential client does not know why he or she fell, reject the case. If the defendant did not create the defect or danger or if the defect or danger is not a common occurrence—a spill on a floor versus a condition that the defendant would probably not notice—avoid the case because it would be based only on the theory of constructive notice. You should still try to prove constructive notice even when it is clear that the defendant created the hazard, but generally do not accept cases based on that theory alone. Seek to prove both the cause of the defect and constructive notice of it.

2. Did the plaintiff suffer serious injuries? If you take cases involving only serious injuries, you can avoid the major jurors' perception that slip-and-fall injuries are minor and that the plaintiffs deserve less compensation than plaintiffs in other cases. For every potential case, use a two-step approach. First, determine whether the injuries are serious enough to hold jurors' attention and evoke their sympathy. If so, you will probably be able to maintain the jury's attention while you tackle the second step—proving liability. If the injuries are not severe, jurors may not consider the fault issues seriously or fairly, as some of them may be biased against both connective-tissue injuries and premises cases.

3. Was the injury immediately reported to the defendant's management? If not, issues arise that could be fatal to the plaintiff's case: his or her credibility may be called into question, and you may lose the opportunity to develop certain evidence against a defendant. For example, how many times have you seen an incident report that documented an employee's verbal admission that the hazard existed at the time of the client's fall? If the plaintiff fails to report the slip and fall at the same time, the potential to use these "excited utterances" to the plaintiff's advantage is lost. An employee may be genuinely affected when the injury happens and may not have time to reflect coolly or be silenced by his or her employer.

4. Is the defendant a corporation or other business entity? Avoid cases against individuals, such as home owners. Most jurors can identify with a home owner, and many reasonable duties of a business owner do not apply to an individual. Corporations, however, spend significant sums on maintenance and safety, and jurors hold businesses to a higher standard of care. Corporate policies and practices often set the standard of care for a business owner. But jurors will remain sympathetic to a home owner/defendant who ignores maintenance.

5. Has the potential client clearly explained the defect or danger? If not, you will have trouble explaining it to the jury. Consider asking an expert to explain the defect or danger. If he or she has difficulty simplifying it, reject the case. Jurors must be able to instantly and easily grasp the liability aspect, or the case is a loser. This is not to say that experts don't play an important role in handling premises liability cases. Their role is explained below.

Call in the Experts

Although many slip-and-fall cases involve safety hazards that can be proven with testimony from lay witnesses or the defendant's employees and management, it is wise to hire an expert witness who can analyze liability issues. The fields of potential expert witnesses range from engineering to aquatic safety to retail merchandising.

Why should you hire an expert? He or she can analyze the defect and danger, determine the applicability of government codes and regulations, and help establish the duty of a reasonable business owner. The liability issues in these cases may be difficult, and the defense is certain to file a motion for summary judgment with supporting affidavits from the defendant's employees. If you have no expert and leave these issues to a judge, you run a substantial risk that your client's case will be dismissed. State law books are full of such cases (1). (Numerals in parentheses in this section and the next refer to the legal notes at the end of each section.)

An expert's analysis and, if necessary, an affidavit establishing the defendant's duty and breach of that duty may help you overcome a motion to dismiss. Without an expert, you have only your client's affidavit and the hope that the trial judge will let a jury hear the case. Don't hope: hire the expert.

Without experts, you must rely on discovery of facts and information that are usually within the sole control of the defendant. Employees are often encouraged to say that they had no notice of a dangerous condition, which helps their employers avoid liability. Further, many corporate defendants in premises liability cases stonewall in discovery to keep critical information out of plaintiffs' hands (2).

An expert can help you establish patterns of consumer behavior that business owners foresee—or should. For instance, a retail expert can explain that stores display the most profitable merchandise at eye level because shoppers are drawn to items displayed there and pay less attention to floor-level items.

Why do warehouse retailers stack merchandise so high when they know that customers generally do not ask for help reaching items? Why do retail employees check store aisles every 15 minutes? They know that items may fall on the floor and that customers may trip over or slip on them. What percentage of customers will spill food at a self-service food bar? Does a building meet safety and building codes?

The answers can help you explain why the defect was not so obvious and how it was created by the business owner.

An expert witness proved indispensable, for example, in a case that our firm handled. A 54-year-old woman had tripped at a restaurant while walking to the restroom. She hit her head and suffered a detached retina, eventually losing 99% of the vision in her dominant eye.

Investigation revealed that the restaurant had been flooded two years earlier and that three floor tiles had been damaged. The tiles were removed but not replaced, leaving a rise in the floor of less than half an inch. The restaurant management had placed a mat over the missing tiles (preventing customers from noticing the change in elevation) and placed a sign warning "Watch Your Step" in the general area.

Discovery of information from restaurant employees showed that other customers and restaurant workers had stumbled in the area but that none had fallen. Was this evidence sufficient to establish liability? Rather than rely on this testimony, we hired a safety engineer who examined the site, photographed and videotaped it, and evaluated applicable building codes. He also prepared a computer-generated animation of our client's trip, stumble, and fall.

The engineer kept his analysis simple. He determined that there was a tripping hazard at the restaurant the day our client had been injured and that reasonable restaurant management should have been aware of it. The expert said that the missing tiles caused the plaintiff to fall, that the flooring defect violated building codes, and that the warning sign was misplaced and insufficient.

Replacing the tiles would have cost only $10, he testified, adding that he had often consulted with major restaurant corporations and that this type of hazard had to be repaired immediately to prevent major, foreseeable injuries to consumers. The analysis appeared straightforward and irrefutable.

The defendant hired its own expert, who concluded that missing tiles did not violate any codes and that the mat had corrected any flooring defect. He also testified that the restaurant management did not need to replace the missing tiles because a sign warning "Watch Your Step" was sufficient. The plaintiff, he said, fell because she had dragged her feet while walking.

The jury concluded that the plaintiff was 10% at fault and the defendant 90% (3).

Although we might have prevailed without an expert, the safety engineer's analysis gave the jury a solid basis for concluding that the defendant had breached its duty of reasonable care and had known of the condition long enough to have corrected it. Without that testimony, the jury would have heard only our legal arguments—and the testimony would have added weight to the verdict if the defense had filed an appeal.

Be aware that some trial judges can be hesitant to let experts testify about premises liability issues that they feel the jury can understand and evaluate themselves. Judges must be persuaded that experts are necessary because duty and breach of duty can no longer be determined by the "reasonable-person" standard that many trial judges have applied in the past (4). Maintaining property and keeping customers safe involves government regulations and codes, engineering principles, and safety factors beyond the knowledge of most laypeople.

Placing the Blame

Typically, the defendant is the best party to prevent a customer from slipping and falling on its premises. Still, the defense that a plaintiff's comparative fault was the legal cause can be effective. Commenting on this defense in a case involving a hazardous sidewalk, a Tennessee court held that "there is a certain amount of care that the Good Lord intended for all of us to take in regard to where we plant our feet, for the world is not flat" (5).

It is possible to overcome the comparative-fault defense. For example, in *Kingsul Theaters, Inc. v. Quillen*, plaintiff's counsel presented the theories of "momentary forgetfulness" and "momentary distraction" to argue that the plaintiff's fault had not caused the fall. The plaintiff successfully argued that she had been momentarily distracted when she turned to look for her daughter. She fell over a six-inch step that she had seen earlier (6).

In some jurisdictions, courts and juries are required to balance the foreseeability and gravity of harm—even if the danger is open and obvious—against the defendant's burden to remedy the harm (7). The Tennessee Supreme Court held that the risk of harm from a deviation in a sidewalk was unreasonable despite its being open and obvious. The court noted that the defendant had actual knowledge of the deviation and that corrective action was available but that it took no steps to warn of the danger. Before courts accepted this balancing test, the plaintiff's fault barred recovery if the danger was open and obvious. Some jurisdictions still follow this rule (8).

The plaintiff's experts must be able to address, from a business standpoint, defenses asserting the openness and obviousness of a hazard and the plaintiff's duty to look where he or she is walking. For example, as noted above, an expert can tell the jury why a store displays merchandise at eye level and how that prevents customers from seeing floor-level dangers.

Without that testimony, these commonsense defense arguments are effective before a judge and jury. Business owners spend millions of dollars on advertising. Often, those ads encourage consumers to visit the premises, and there are few warnings that the consumer should anticipate or expect dangers.

Consumers injured on business property need representation. By applying strong screening criteria, using qualified experts, and preparing to meet the comparative-fault defense, you can hold businesses accountable for failing to provide safe premises (John D. McMahan, Trial).

Notes

1. See, e.g., *Bailes v. Public Bldg. Auth.*, No. 03A01-9605-CV-00157, 1996 WL 722042 (Tenn. Ct. App. Dec. 17, 1996); *Rice v. Knoxville Utils. Bd.*, No. 03A01-9606-CV-00209, 1996 WL 732477 (Tenn. Ct. App. Dec. 23, 1996); *Tracy v. Exxon Corp.*, No. 02401-9512-CV-00277, 1996 WL 741876 (Tenn. Ct. App. Dec. 31, 1996).

2. See Gilbert T. Adams III & Alto v. Watson III, *Big Box Retailers: Discovery Abuse*, TRIAL, Apr. 2000, at 38; see also Bruce S. Kramer and Elaine Sheng, *Busting Open the Big Box*, on page 26 of this issue.

3. *Sanders v. Jenkins Rest. & Deli*, No. V-99-716 (Tenn., Bradley County Cir. Ct. Jan. 24, 2001).

4. See Clifford Britt, *Getting Your Security Expert Over the Daubert Hurdle*, on page 31 of this issue.

5. *Coln v. City of Savannah*, No. 02A01-9507-CV-00152, 1996 WL 544652 at * 4 (Tenn. Ct. App. Sept. 25, 1996), rev'd by 966 S.W.2d 34 (Tenn. 1998).

6. 196 S.W.2d 316 (Tenn. Ct. App. 1946). See also *Keller v. Vermeer Mfg.*, 360 N.W.2d 502 (N.D. 1984); *Soileau v. S. Cent. Bell Tel. Co.*, 406 So. 2d 182 (La. 1981).

7. See, e.g., *Coln v. City of Savannah*, 966 S.W.2d 34.

8. See *Bennett v. Stanley*, 748 N.E.2d 41 (Ohio 2001); *Dowen v. Hall*, 548 N.E.2d 346 (Ill. App. Ct. 1989).

ALTERNATIVE LIABILITY THEORIES IN SLIP-AND-FALL CASES (FROM EHOW.COM, WRITTEN BY JACK HICKEY)

Trial: April 1, 2000
Attorney: Jack Hickey

Theories of negligent maintenance and negligent method of operation may be used to show that a defendant is responsible for a client's injuries.

A potential client enters your office and tells you that she broke her leg when she fell at the ABC supermarket. You begin screening the viability of the case by asking what caused the fall.

If you're familiar with slip-and-fall cases, you know that the client could respond with one of several replies: "I don't know," "I fell in this huge puddle of water that looked like it was coming out from under a refrigerated case where they store the meats," or "I fell on some trash. I think that it was a paper cup."

If her answer points to water or trash, you ask, "How long was it on the floor?" More often than not, the client will tell you that she has no idea. Don't despair. In some cases, time is not an issue. For example, if the water was streaming from a refrigerated case, the defendant store was responsible for putting the water on the floor. It doesn't matter how long it was there.

However, in some cases—such as when a customer rather than a store employee caused the hazardous condition—time is an issue. In these cases, the plaintiff has to prove that the defendant had actual or constructive knowledge of the dangerous condition. Constructive knowledge may be inferred from either the amount of time a substance has been on the floor or the fact that the condition occurred with such frequency that the owner should have known about it (1).

When you cannot show how long a substance was on the floor or how it got there, you can prove your client's case using alternative liability theories based on negligent method of operation or negligent maintenance.

These theories provide a basis for liability either where the defendant, through its repeated conduct, has created a dangerous condition (negligent method of operation) or where the defendant has not cleaned up a dangerous condition, which may not have been caused by the defendant (negligent maintenance).

The court should recognize both theories as separate from the requirements of actual or constructive notice (2). The key issue is foreseeability: did the defendant, given the circumstances, have a reasonable opportunity to foresee the accident?

Courts have held that the negligent method of operation theory applies in the following instances:

- A racetrack sold bottled drinks and told its patrons to leave the empty bottles anywhere (3).
- A grocery store used ice to chill produce, and when the ice melted, the water leaked onto the floor (4).
- A grocery store served meat samples to customers, some of whom dropped them on the floor; (5)
- A grocery store chain sold greens from self-serve bins, and customers dropped the vegetables on the floor (6).
- A store sold avocado juice and allowed customers to consume the beverage as they shopped, resulting in spills (7).
- A discount department store chain allowed a McDonald's to sell food in an area adjoining the store (8).

Some courts use a two-step analysis to determine whether the negligent method of operation theory applies. First, the plaintiff must prove that the method of operation is inherently dangerous or that the operation is being conducted in a negligent manner. Second, the plaintiff must show that the condition of the floor was created as a result of the negligence.

A word of caution: Some courts have held that the negligent method of operation theory applies only to a specific type of facility, such as an arena where masses of people are expected. These rulings seem to specifically exclude supermarkets and make distinctions without differences. There is no real reason to carve out most cases from this rule, but some courts seem anxious to do so (9).

In a recent case in which judgment for the plaintiff was affirmed, the appellate court held that evidence that a trash can was repeatedly overflowing was admissible to prove foreseeability and that a condition occurring with sufficient "regularity" is an alternative to proving that the defendant had notice (10).

The theory of negligent maintenance can be used where the dangerous condition—even if caused by an outside source—either occurs often or is allowed by the defendant to occur. For example, the defendant store owner may be guilty of negligent maintenance when an area of the floor is constantly getting wet from a source that the defendant does not control and the defendant fails to clean it up. A customer should be able to bring a claim without proving how long the substance was on the floor.

One court articulated this theory in a case involving a cruise ship when it held that

> actual or constructive knowledge is irrelevant in cases not involving transitory, foreign substances (i.e., the typical banana peel case), if ample evidence of negligent maintenance can be shown. Here, plaintiffs offered testimony that the constant wetness on the deck, even on otherwise dry days, could be attributable to sea spray or to condensation caused by steam escaping from a nearby smokestack, but that nevertheless, the surface of the deck easily could have been made skid-proof if properly coated and maintained. (11)

The premises owner should not escape liability for a problem that occurs repeatedly merely by posting a sign in the area. Signs can be too little, too late, certainly for someone walking through an area inundated by signage, lighting, people, noises, and other distractions. In fact, warning signs can be used as evidence that the premises owner knew of the dangerous condition (12).

Potential Defendants

Potential defendants in any premises liability case include the owner and manager of the property. Most courts hold that ownership alone is not sufficient to extend liability to a party. The owner must have had some management authority to control the property (13).

To determine the identity of the owner, check the tax rolls and real property records. After you get the owner's name, call or write the owner to determine whether some other entity managed the property. You should also visit the premises to see whose name appear on the signs posted there.

The client's receipts from the store, hotel, or other premises may also identify the management company. The phone book is another source of information. The listing for the premises may identify the corporation or other business entity that manages it.

Other potential defendants include food service concessionaires and cleaning service providers. These are typically independent contractors of the business owners or managers.

The concessionaire in a concert hall, for example, may be liable under the negligent method of operation theory because it failed to safely transport food from the kitchen to the concession and food spilled out of the containers and caused a slip and fall. The concessionaire may also be liable under a negligent maintenance theory if it allows restocking of the stand to occur during a concert.

Independent cleaning service companies may be responsible for accidents in retail stores or commercial office buildings. A cleaning service may leave wax or polish residues on the floor, causing falls. Or cleaning crews may begin mopping floors before workers have left office buildings. The service should not only post warning cones by slippery floors but also go into offices or shopping areas to warn people verbally of a hazard. Otherwise, people may walk onto a slick floor before they see the warning cones.

Franchisors are another group of potential defendants. Although they will claim that their franchise agreement stipulates that they have no connection to the business's management, the test is not whether the company calls itself a franchisor, owner, or manager or uses any other title. The test is whether the company has control or the right to control the management of the premises.

Franchise documents may reveal that the franchisor sets regulations and policies about managing the premises. The franchisor may also have the right to periodically inspect the premises to verify that the franchisee's management has met the standards.

Franchisors usually have the right to terminate a franchise or warn a franchisee that it must bring itself into compliance. This is control, which could subject the franchisor to liability.

Discovery

The initial request for documents should include, at a minimum, a request for agreements between the business owner and the property management company, the property management company and its franchisor, and the property management company and its subcontracted cleaning company.

These agreements will help identify defendants based on the amount of control the entities have over cleaning, maintenance, repair, and safety operations of the premises. The agreements may also delineate the companies' responsibilities for maintenance of cleaning. These documents can often be useful in a deposition in which company managers are unaware of their defined responsibilities.

Manuals and handbooks on operations, safety, loss prevention, and cleaning also can be essential. Handbooks may prescribe cleaning methods, violations of which can be evidence of negligence. Handbooks also may articulate that mopping or allowing substances to remain on the floor may cause injury. Have your experts review these manuals, express the industry standard, and provide an opinion that the defendant should have complied with the standard and, by not complying, violated that standard. If the defendant has no manuals, this fact may be useful in establishing negligence.

Also obtain employee videos regarding training, safety, accident prevention, cleaning, and maintenance. These will describe risks inherent in the business that the manager may not readily admit. The videos may provide information you can use to develop deposition questions for the business manager.

In addition, you should obtain employee time records to determine who was on duty at the time of the accident. A manager may testify that three people clean the area at all times when time sheets show that only two people worked that day. The employee roster will also provide subjects to depose.

Experts

Experts may not be necessary in these cases, but they can be helpful. In one negligent method of operations case involving a concert hall, my firm retained a former dean of a local university school of hospitality. We showed that before or during concerts, the concessionaire restocked a concession stand with Coca-Cola syrup by hand-carrying boxes of the syrup from a storage room to the concession stand about 300 feet away. The expert testified that the concessionaire should have used different procedures to restock the stands. The restocking should have taken place before the event, and the food and beverages should have been transported in a cart with sides to contain spills (14).

The same expert testified that both the concessionaire and the management company used negligent cleaning procedures. The management company did not assign cleaners to any zones in the concert hall and did not require them to inspect the floor in any particular pattern or route. The cleaners conceivably could have walked through and inspected only certain areas, leaving others uninspected and uncleaned. Also, the expert testified that cleaners who were not assigned particular zones or routes would travel routes that were out of their supervisor's view, then congregate and begin socializing with each other.

The theories of negligent method of operation and negligent maintenance are not new, but they tend to be overlooked by attorneys handling slip-and-fall cases. They are based on the age-old concept of foreseeability. When more common liability theories won't work, these alternatives can help you pursue your clients' claims (Jack Hickey, Trial).

Notes

1. *Publix Super Market, Inc. v. Sanchez*, 700 So. 2d 405, 406 (Fla. 3d Dist. Ct. App. 1997) (per curiam), review denied, 717 So. 2d 537 (Fla. 1998); *Schaap v. Publix Supermarkets, Inc.*, 579 So. 2d 831, 834 (Fla. 1st Dist. Ct. App. 1991).

2. See, e.g., *Wal-Mart Stores, Inc. v. Reggie*, 714 So. 2d 601 (Fla. 4th Dist. Ct. App. 1998). *Sanchez*, 700 So. 2d 405, 406; *Mabrey v. Carnival Cruise Lines, Inc.*, 438 So. 2d 937, 938 (Fla. 3d Dist. Ct. App. 1983).

3. *Wells v. Palm Beach Kennel Club*, 35 So. 2d 720, 721 (Fla. 1948).

4. *Torda v. Grand Union Co.*, 157 A.2d 133, 135 (N.J. Super. Ct. App. Div. 1959).

5. *Little v. Butner*, 348 P.2d 1022 (Kan. 1960).

6. *Wollerman v. Grand Union Stores, Inc.*, 221 A.2d 513, 514 (N.J. 1966).

7. *Jackson v. K-Mart Corp.*, 840 P.2d 463, 466-67 (Kan. 1992).

8. *Gump v. Wal-Mart Stores, Inc.*, No. 21670, 1999 Haw. App. LEXIS 180, at *30 (Haw. Ct. App. Nov. 17, 1999).

9. See, e.g., *Soriano v. B&B Cash Grocery Stores, Inc.*, No. 98-1668, 1999 Fla. App. LEXIS 5721, at *5–*6 (Fla. Dist. Ct. App. May 5), review granted, 744 So. 2d 456 (Fla. 1999); *Rowe v. Winn-Dixie Stores, Inc.*, 714 So. 2d 1180, 1181 (Fla. 1st Dist. Ct. App. 1998), review denied, 731 So. 2d. 650 (Fla. 1999); *Schaap*, 579 So. 2d 831.

10. See *Wal-Mart Stores, Inc. v. Reggie*, 714 So. 2d 601 (Fla. Dist. Ct. App. 1998).

11. See *Mabrey v. Carnival Cruise Lines, Inc.*, 438 So. 2d 937, 938 (Fla. 3d Dist. Ct. App. 1983) [citation omitted].

12. See, e.g., *id.* (holding that a "slippery when wet" sign was evidence that the defendant knew the deck was dangerous).

13. See, e.g., *Craig v. Gate Maritime Properties, Inc.*, 631 So. 2d 375, 377 (Fla. Dist. Ct. App. 1994); *Regency Lake Apartments Assoc. Ltd. v. French*, 590 So. 2d 970, 974 (Fla. 1st Dist. Ct. App. 1991).

14. *Sullivan v. CDC/SMT, Inc.*, No. 96-013788-CA-03 (Fla., Broward County Cir. Ct. settled Nov. 12, 1998).

III

THE SOLUTION

Codes, Standards, and Regulations: What's It All About?

THE U.S. NATIONAL STANDARDS STRATEGY

The acceptance that there are many ways of developing codes and standards is reemphasized in the U.S. National Standards Strategy (NSS) strategic approach to standards. The strategy was developed by the private sector under the leadership of the American National Standards Institute (ANSI) and has moved standards development away from governmental departments such as the National Bureau of Standards to that of the private-sector standards-developing organizations like ANSI.

The NSS lays out the principles for developing national or international standards to meet society's needs. "The sector-based approach allows interested parties to address their own issues and develop working methods that fit the problems at hand, since no single standardization approach can satisfy all needs." The NSS recognizes that there is "no simple recipe that can be handed down to fit all needs."

OFFICE OF MANAGEMENT AND BUDGET CIRCULAR A119 AND PUBLIC LAW 104-113

Before 1996, the Office of Management and Budget (OMB) Circular A-119 was an advisory that basically recommended the federal agencies to use nongovernmental standards. In 1996, however, the circular was codified by the National Technology Transfer and Advancement Act of 1995.

The codification of the circular made it mandatory for agencies to use private voluntary standards to the extent feasible. If it is not practicable and the agency must rely on government standards, the agency is required to explain why. Examples may include standards for special services or products where the Department of Defense for security reasons may opt to use military specifications.

The National Institute of Standards and Technology (NIST) is the chair of the Interagency Committee on Standards Policy. It is NIST's responsibility to report annually to OMB on the progress the federal agencies are making toward using voluntary standards rather than agency-unique standards.

Neither OMB A-110 nor the Public Law (PL) 104-113 reference ANSI accreditation as a requisite under the definition of the term "voluntary consensus standard" or "voluntary consensus standards body." Further, the federal government will not and cannot show favoritism toward any one organization, whether ANSI is accredited or not.

In a letter addressed to the International Code Council (ICC) dated February 11, 2002, NIST clarified this question by referring to OMB Circular A-119: "In section 4.a.1, the circular states that: 'a voluntary consensus standards body is defined by the following attributes: (1) Openness; (2) Balance of interest; (3) Due process; (4) An appeals process; and (5) Consensus, which is defined as general agreement, but not necessarily unanimity.' In addition, in second 6.h, the circular states: 'This policy does not establish a preference among standards developed in the private sector. Consequently, neither OMB nor NIST can endorse or recognize one standards-developing organization as preferable to another.'"

Truth and objectivity should be the crux of the debate, and the debate should focus on the merits of the codes developed under any consensus process that conforms to the principles outlined in the NIST letter referred to in this article (Yerkes 2003).

> The absence of unified safety standards has served to increase the growing number of slips, trips, and falls.

Is there a standard that defines how slippery a floor can be? Sort of. To date, the ASTM has yet to publish a standard that would define the safety threshold for hard-surface floors. Although the ASTM D-2047 and ASTM C-1028 standards describe how to measure the slip resistance of specific categories of flooring, neither standard defines a margin of risk or safety. However, with the recent publication of the ANSI/NFSI B101.1 standard in 2009, the U.S. now has its first unified floor safety standard that defines the slip risks associated with hard-surface floors.

This is particularly refreshing since this new standard comes on the coattails of a shrinking of the Occupational Safety and Health Administration (OSHA). Over the past decade, both OSHA and the Americans with Disabilities Act (ADA) actually made negative progress on the subject of walkway safety. For those who say that they are in compliance with the ADA or OSHA standard for walkway safety, I say, "Really?" That would be difficult since neither agency actually has a published standard defining an acceptable level of slip resistance, let alone safety. In fact, this lack of clarity has led many property owners to think that they were in compliance with a policy that was nonexistent.

Many safety professionals believe that OSHA and the ADA act as standards developers when in fact they don't (at least when it comes to walking and working surfaces). What both agencies do is rely on the ASTM or ANSI to publish such standards to which they cite.

The fact that OSHA does not have a published slip-resistance standard hasn't prevented OSHA inspectors from citing property owners for their slippery floors. Under OSHA's general duty clause, an OSHA inspector can cite a property owner for having floors that are too slippery, claiming that a particular walkway is "very slippery when wet causing a slipping hazard to the employees" and requiring that a business owner "immediately investigate the alleged conditions and make any necessary corrections or modifications." The investigator's citation will also require that the property owner provide the OSHA inspector with the results of their investigation as well as any supporting documentation of their findings, including any applicable measurements or monitoring results, photographs or videos, and a description of any corrective action the property owner has taken or is in the process of taking, including of the corrected condition. The property owner is provided a series of forms that he or she generally is given five days to complete. Needless to say, once a property owner appears on OSHA's "radar screen," he or she tends to stay there.

STATE, COUNTY, OR CITY ORDINANCES

It is possible for states, cities, and counties to establish their own standards or ordinances regarding issues of public safety. The City of Los Angeles established what it calls a "maintenance covenant," which requires property owners "to comply with the slip-resistant ground and floor surface requirements contained in Part 2 of Title 24 of the California Code of Regulations relative to accessibility for all ground and floor surfaces including floors, walks, ramps, stairs, and curb ramps in buildings located within the city's borders."

The city's covenant requires property owners "to maintain or require maintenance by tenants of said building of all ground and floor surfaces in said building as slip-resistant. Where ground and floor surfaces are not inherently slip-resistant, such surfaces shall be made slip-resistant by either permanent etching of the surface or by application of a surface treatment including the periodic re-application of such treatment in accordance with the manufacturer's specifications. Surface treatment shall meet the requirement for slip-resistance, which can be accomplished in accordance with either a product label or manufacturer's specification indicating that the surface treatment meets an ASTM standard for slip-resistance for the ground and floor surfaces being treated or by having the treated surface tested by a City of Los Angeles–

approved testing laboratory in accordance with an ASTM standard for slip-resistance. Ground and floor surfaces shall be considered slip-resistant if the static coefficient of friction measured for such surfaces is a minimum of 0.8 for ramps or a minimum of 0.6 for other accessible routes when tested in accordance with an ASTM standard for slip-resistance."

The only problem with the covenant agreement is that it requires property owners to "use chemical treatments that meet the ASTM standard for slip resistance," while the ASTM has yet to publish a standard for such products. The only nationally recognized standard that rates the performance of a chemical safety treatment is that of the 101-B standard of the National Floor Safety Institute (NFSI), which is discussed later in the book.

In today's debate over model codes and standards, we have begun to hear negative publicity about the ICC codes because they lack accreditation by ANSI. Many safety experts, on the other hand, believe that only ASTM standards should be referenced. It has been the position of the federal government to "require" its agencies to recognize only ANSI-accredited codes and standards, leaving the ASTM in the cold. Understanding this debate requires a little background in the sometimes arcane language of the codes and standards world.

THE ICC

The ICC, a membership association dedicated to building safety and fire prevention, develops the codes used to construct residential and commercial buildings, including homes and schools. Most U.S. cities, counties, and states that adopt codes choose the international codes developed by the ICC. The vision and mission of the ICC can be summarized as follows:

- *Vision.* Protecting the health, safety, and welfare of people by creating better buildings and safer communities
- *Mission.* Providing the highest-quality codes, standards, products, and services for all concerned with the safety and performance of the built environment

The ICC was established in 1994 as a nonprofit organization dedicated to developing a single set of comprehensive and coordinated national model construction codes. The founders of the ICC are Building Officials and Code Administrators International, Inc. (BOCA); the International Conference of Building Officials; and Southern Building Code Congress International, Inc. Since the early part of the twentieth century, these nonprofit organizations developed the three separate sets of model codes used throughout the United States. Although regional code development has been effective and responsive to our country's needs, the time came for a single set of codes. The nation's three model code groups responded by creating the ICC and by developing codes without the regional limitations imposed by international codes.

There are substantial advantages in combining the efforts of the existing code organizations to produce a single set of codes. Code enforcement officials, architects, engineers, designers, and contractors can now work with a consistent set of requirements throughout the United States. Manufacturers can put their efforts into research and development rather than designing to three different sets of standards and can focus on being more competitive in worldwide markets. Uniform education and certification programs can be used internationally. A single set of codes may encourage states and localities that currently write their own codes or amend the model codes to begin adopting the international codes without technical amendments. This uniform adoption would lead to consistent code enforcement and higher-quality construction. The code organizations can now direct their collective energies toward wider code adoption, better code enforcement, and enhanced membership services. All issues and concerns of a regulatory nature now have a single forum for discussion, consideration, and resolution. Whether the concern is disaster mitigation, energy conservation, accessibility, innovative technology, or fire protection, the ICC provides a single forum for national and international attention and focus to address these concerns.

CODES

Let's begin with the term "codes." Codes are systematically arranged laws and regulations. The model building codes are not considered legal documents until they are adopted by a governmental jurisdiction like that of a state, county, or municipality. Codes are developed to be accepted into local or state laws, but as "models" they are developed in much the same manner that state or local governments are required to develop laws.

The ICC follows a process similar to that used by state or local governments to develop and maintain its family of model codes, known as the international codes. The ICC refers to its process as a governmental consensus process. It is an open, inclusive, and balanced consensus process with built-in safeguards to prevent domination by any single vested interest. The system ensures fairness in the process, controls against conflicts of interest, and prevents vested economic interest from determining the outcome of the final vote.

The enforcement of codes falls largely on state and local governments in the United States. This enforcement authority is derived from the Tenth Amendment to the U.S. Constitution, which "reserves to the states" the right to legislate for all matters not delegated to the federal government or prohibited to the states by the federal government, including laws for the protection of the public's health, safety, and welfare.

In the 1987 version of the BOCA code, there are only three brief references to slip resistance of walkways, stairs, and ramps, totaling less than a page, while there are at least 67 pages of a total of 477 pages devoted directly to fire safety, and many more—such as the sections on egress—that are indirectly related to fire safety. Of the three cited references to "slip-resistant surfaces," none actually define or specify the criteria for measuring slip resistance. Thus, any meaningful use of these provisions requires reference to other sources, such as ASTM or ANSI standards, none of which are referenced in the BOCA code.

STANDARDS, RECOMMENDED PRACTICES, AND GUIDELINES

Voluntary standards, recommended practices, and guidelines are not written in mandatory language. The term "standard" means a common and repeated use of rules, conditions, guidelines, or characteristics for products or related processes, production methods, and related management system practices. Standards are often referred to in codes, but without codes to place them in an overall enforcement context, their impact on building construction lacks a systematic approach to protection of lives and property. The term "standard" may take on several different meanings. Each industry may have an educational or professional organization that produces "standards" or what are often referred to as "standard practices" for their specific industry. For example, the International Sanitary Supply Association has published a consensus-based Cleaning Industry Management Standard (CIMS), which janitorial maintenance companies are encouraged to practice. The CIMS is not the same as an ASTM or ANSI standard but can fill the gaps in the absence of such a standard.

ASTM INTERNATIONAL

Formed more than a century ago and originally known as the American Society for Testing and Materials (ASTM), ASTM International is a nonprofit organization that provides a forum for producers, independent researchers, and general-interest participants to meet and write voluntary standards for materials, products, medical services and devices, computer systems, and industry.

As one of the largest voluntary standards-developing organizations in the world, ASTM International has earned a reputation as a trusted source for technical standards for materials, products, systems, and services. Standards developed at ASTM are the work of more than 30,000 ASTM members and 132 standards-writing committees that produce more than 10,000 standards each year. Participation in ASTM International is open to all with a material interest, anywhere in the world.

Four ASTM committees govern walkway slip-resistance issues:

ASTM D-21—finishes and polishes

ASTM C-21—ceramic whitewares and related products

ASTM F-06—resilient floor coverings

ASTM F-13—safety and traction for footwear

Listed here are the current standards that pertain to the measurement of walkway slip resistance as published by the ASTM:

ASTM F-1637—"Standard Practice for Safe Walking Surfaces"

ASTM D-2047—"Standard Test Method for Static Coefficient of Friction of Polish-Coated Flooring Surfaces as Measured by the James Machine"

ASTM C-1028—"Standard Test Method for Determining the Static Coefficient of Friction of Ceramic Tile and Other Like Surfaces by the Horizontal Dynamometer Pull Meter Method"

ASTM F-609—"Standard Test Method for Using a Horizontal Pull Slipmeter (HPS)"

Two popular ASTM standards dealing with the measurement of walkway slip resistance were introduced in 1996 and officially withdrawn in 2006. They are ASTM F-1677 (Brungraber PIAST) and ASTM F-1679 (English XL VIT). Both standards were withdrawn for two reasons. First, neither of the two tribometers named in the standards had satisfactorily provided consistent results. Second, because both of the tribometers named in each standard were patented, they were in direct conflict with the ASTM's "Regulations," to which each standard did not provide for an "alternative propriety method for measuring slip."

ANSI

Founded in 1918 as "the voice of the U.S. standards and conformity assessment system," ANSI is comprised of more than 125,000 companies and 3.5 million professionals, including government agencies, organizations, companies, academic and international bodies, and individuals. ANSI's goal is to "strengthen the U.S. marketplace position in the global economy while helping to assure the safety and health of consumers and the protection of the environment."

ANSI oversees the creation, promulgation, and use of thousands of norms and guidelines that directly impact businesses in nearly every industry sector. ANSI is also actively engaged in accrediting programs that assess conformance to standards—including globally recognized cross-sector programs such as the International Organization for Standardization (ISO) 9000 (quality) and ISO 14000 (environmental) management systems.

The mission of ANSI is "to enhance both the global competitiveness of U.S. business and the U.S. quality of life by promoting and facilitating voluntary consensus standards and conformity assessment systems, and safeguarding their integrity."

ANSI provides third-party accreditation to approximately 200 standards developers in the United States. ANSI was created to ensure that voluntary consensus standards for products, processes, and services are developed with integrity.

The ICC, the NFSI, the American Society of Safety Engineers (ASSE), and leading organizations have been recognized by ANSI and awarded the distinction of being named an accredited standards-developing organization (ASD). ANSI has signed an agreement to distribute the international codes and also distributes other non–ANSI-accredited standards through its electronic store. The two ANSI committees that address slips, trips, and falls are the ANSI/NFSI B101 Committee on the Prevention of Slips, Trips, and Falls. ANSI/NFSI B101 represents the leading edge of slip, trip, and fall prevention standards and is currently developing eight slip, trip, and fall prevention standards. The other committee is the ANSI/ASSE A1264 committee, which authored the A1264.2 standard for workplace safety.

ANSI/NFSI B101.1-2009

The ANSI/NFSI B101.1-2009 standard, which had gone through four iterations by the NFSI, was further developed by a subcommittee of the NFSI B101 Standards Committee on Slip, Trip, and Fall Prevention. National in scope and governed by the procedures of ANSI with the NFSI as the ANSI ASD, this standard establishes a test method that specifies the procedures and devices used for both laboratory and field testing to measure the wet static coefficient of friction (SCOF) of common hard-surface floor materials.

The B101.1 standard was originally published as a test method by the NFSI in 2002 under the title NFSI 101-A and has served as the basis of materials testing and product certification under the NFSI's product certification program. It was the intent of the NFSI to develop a voluntary test method whose purpose was to establish a uniform test method for measuring the wet SCOF of floor coverings, polishes, and walkway coatings.

It is intended that the procedures and performance requirements contained in the ANSI/NFSI B101.1-2009 standard will be adopted by affected professionals and property owners as the measurement procedure for determining traction levels that facilitate the remediation of walkway surfaces when warranted.

The ANSI/NFSI B101.1-2009 standard establishes three "traction ranges" that are intended to assist property owners in better understanding the real-world slip-and-fall risk associated with their walkways. In the past, walkway safety was seen as meeting a single-value threshold (i.e., ASTM D-2047 dry SCOF value of 0.5 or greater) whereby the walkway surface either met this value or didn't and either "passed" or "failed." Users of such standards would logically infer that floors were either "good" or "bad" or "safe" or "unsafe," providing property owners with little insight as to the real-world risk potential for a slip and fall.

The ANSI/NFSI B101.1-2009 standard has established a new approach to walkway safety. Floors are seen no longer as safe or unsafe but rather as having a defined risk level. Walkways whose wet SCOF are high logically provide a lower level of risk for a slip and fall, while walkways that possess a low wet SCOF increase the risk of a slip and fall. Listed in table 8.1 are the three published traction ranges defined in the ANSI/NFSI B101.1-2009 standard.

NFSI

The NFSI was founded in 1997 as a 501(c) (3) not-for-profit organization with a mission to educate business and property owners as an aid in the prevention of slips and falls through education, training, and research. Since 1997, the NFSI has become the leading authority in slip, trip, and fall prevention and in 2006 was awarded the distinction of being an ANSI ASD.

The mission of the NFSI is "to aid in the prevention of slips, trips, and falls through education, research, and standards development."

The NFSI is a membership-based organization that provides a wide range of publications and training in the field of walkway safety. The NFSI also provides product testing and "certifies" flooring materials, coatings, chemical floor-cleaning products,

Table 8.1. The Three Published Traction Ranges Defined in the ANSI/NFSI B101.1-2009 Standard

Wet SCOF Value (μ)	Available Traction	Remediation
μ ≥ 0.60	High: lower probability of slipping	Monitor SCOF regularly and maintain cleanliness
0.40 ≤ μ < 0.60	Moderate: increased probability of slipping	Monitor SCOF regularly and maintain cleanliness; consider traction-enhancing products and technologies
μ < 0.40	Minimal available: higher probability of slipping	Seek professional intervention; consider replacing flooring and/or coating with high-traction products

and treatments. NFSI-certified high-traction products are evaluated per either the NFSI 101-A or the NFSI 101-B standard and are frequently specified by consumers, property owners, and insurance companies nationwide.

The NFSI also provides a wide range of classroom training programs, including the Walkway Auditor Certification (WAC) program. Safety professionals, product manufacturers, and floor maintenance professionals who seek to enter the field of walkway safety can become certified walkway auditors through a multi-day classroom training program. The NFSI WAC is the most respected professional training course of its kind and serves as the model program for individuals working in the safety or legal industries (e.g., expert witnesses).

NFSI 101-A Standard for Measuring Wet SCOF of Common Walkway Materials and Coatings

The purpose of the NFSI B101-A standard is to specify the test method for measuring the wet SCOF of common hard-surface flooring material (i.e., ceramic floor tile, vinyl floor coverings, wood laminates, and floor coatings and polishes) and employs the use of an NFSI-approved tribometer as specified by the tribometer selection process. This test method does not apply to mechanically polished tile (e.g., polished porcelain or marble) or carpeting of any type. Walkways that meet a minimum wet SCOF of 0.6 or greater are classified as high-traction surfaces.

NFSI 101-B Standard for Measuring Wet SCOF of Chemical Floor-Cleaning Agents and Treatments

The NFSI 101-B standard is identical to that of the NFSI 101-A standard in that it seeks to measure the wet SCOF of a surface; however, the NFSI 101-B standard applies only to walkway surfaces that have been cleaned with a chemical floor-cleaning agent or chemical treatment. Products that increase the wet SCOF by 20% or more are classified as high traction. The NFSI certifies products as high traction and, depending on the product, uses either the NFSI 101-A or the NFSI 101-B standard.

NFSI 101-C Test Method for Measuring Dry TCOF of Floor Mat Backing Materials

Floor mat migration (movement) is a leading cause of both commercial and residential trip and fall injuries. In April 2010, the NFSI released its NFSI 101-C "High-Traction" floor mat standard. This new test method specifies the procedures and devices used for laboratory testing for the measurement of the dry Transitional Coefficient of Friction (TCOF) of floor mat backing materials. The NFSI 101-C test method provides both the measurement procedures and traction requirements for floor mat backings to be considered "high-traction" and serves as a useful tool in providing product manufacturers and their customers a standard by which to determine the slip resistant qualities of floor mats, rugs, and runners.

PSEUDO–STANDARDS-DEVELOPING ORGANIZATIONS

The Consumer Product Safety Commission

This small and often overlooked department of the federal government is not a standards-developing organization, but it does an excellent job of identifying consumer-related product risk and in turn notifying the consumer as to possible safety-related risks. Technically, products such as floor coverings, floor care products, and footwear fall under the umbrella of the Consumer Product Safety Commission (CPSC) but as of yet have not been identified as big enough problems requiring their attention. Their failure to do so has more to do with their size and budget rather than their passion or focus on consumer safety. The CPSC has a talented and well-trained staff, including safety engineers, ergonomists, and other safety and health professionals. They simply have not made slip, trip, and fall prevention a priority.

Centers for Disease Control

Like the CPSC, the Centers for Disease Control (CDC) is a relatively small government agency whose budget simply does not permit it to tackle all the problems affecting the safety of Americans. The leadership and staff of the CDC are highly professional, focused, and passionate about their work. The CPSC is very much aware of the problem of falls, especially as they apply to the nation's elderly population. In 2006, the CPSC signed a memorandum of understanding with the NFSI calling for the two organizations to jointly address "the growing epidemic of slip and fall accidents." The CDC does not author walkway safety standards but has a big influence on the health and welfare of Americans.

The National Safety Council

On August 13, 1953, President Dwight D. Eisenhower signed a bill granting the National Safety Council (NSC) a federal charter. With the stroke of his pen, the NSC was elevated to a federal corporation. This act did not mean that the council became a government agency or would receive federal subsidies. Instead, it "bestowed" the prestige of official recognition, then–council president Ned H. Dearborn wrote. In accepting the charter from Davis, Colonel John Stilwell, vice chairman of the council's board of trustees, said that the government had given the council "not only the right but the obligation" to continue leading "a relentless and unceasing fight to reduce the accident toll in every field of human activity—in the home, on the job, in traffic, on the farm, among children—everywhere and every place that accidents occur."

To many, the green-cross logo representing the NSC is seen as the authority in the safety field, and it is. However, the NSC is not a standards-developing organization. The promotion of safety by the NSC is done primarily through its more than 40,000 members, most of which are companies. The NSC has served as a champion of safety in multiple areas of safety; however, slip, trip, and fall prevention has not been one of them. In the past, the NSC's leadership has chosen to focus their resources on other areas of accident prevention, and many safety professionals have found this surprising since the NSC's own "accident facts" data have revealed that slips, trips, and falls are the leading cause of accidental injury facing the country.

The Home Safety Council

The Home Safety Council (HSC) was an outgrowth of Lowe's Companies, Inc., and to this day, Lowe's is its biggest source of funding. The HSC's emphasis is, as its name states, home safety. Although 60% of all falls take place in the home, the HSC spends little time researching the subject of slips, trips, and falls and provides only cursory information (e.g., safety checklists) to its members.

WHAT OSHA HAS TO SAY ABOUT WALKWAY SAFETY

The Department of Labor (OSHA) Code of Federal Regulations 29 Section 1910.22 General Requirements applies to all permanent places of employment, except where only domestic, mining, or agricultural work is performed. Measures for the control of toxic materials are considered to be outside the scope of this section:

1910.22(a) "Housekeeping."
1910.22(a)(1)
All places of employment, passageways, storerooms, and service rooms shall be kept clean and orderly and in a sanitary condition.

1910.22(a)(2)
The floor of every workroom shall be maintained in a clean and, so far as possible, a dry condition. Where wet processes are used, drainage shall be maintained, and false floors, platforms, mats, or other dry standing places should be provided where practicable.

1910.22(a)(3)
To facilitate cleaning, every floor, working place, and passageway shall be kept free from protruding nails, splinters, holes, or loose boards.

1910.22(b) "Aisles and passageways."

1910.22(b)(1)
Where mechanical handling equipment is used, sufficient safe clearances shall be allowed for aisles, at loading docks, through doorways and wherever turns or passage must be made. Aisles and passageways shall be kept clear and in good repairs, with no obstruction across or in aisles that could create a hazard.

1910.22(b)(2)
Permanent aisles and passageways shall be appropriately marked.

OSHA's proposed rule goes a bit further by expanding the following sections; however, as of the publication of this book, the sections listed below have yet to be adopted:

(d) Inspection, maintenance, and repair. (1) The employer shall ensure through regular and periodic inspection and maintenance that walking and working surfaces are in safe condition for employee use.

(3) Only qualified persons shall be permitted to inspect, maintain, or repair walking and working surfaces except for the incidental cleanup of non-toxic materials.

APPENDIX A TO SUBPART D—COMPLIANCE GUIDELINES

Slip Resistance: A reasonable measure of slip-resistance is static coefficient of friction (COF). A COF of 0.5, which is based upon studies by the University of Michigan and reported in "Work Surface Friction: Definitions, Laboratory and Field Measurements, and a Comprehensive Bibliography," is recommended as a guide to achieve proper slip-resistance. A COF of 0.5 is not intended to be an absolute standard value. A higher COF may be necessary for certain work tasks, such as carrying objects, pushing or pulling objects, or walking up and down ramps (U.S. Department of Labor, OSHA).

Question: What is OSHA's COF requirement for walkways?
Answer: OSHA does not have a COF requirement.

Listed below is the letter of interpretation dated March 21, 2003, from Noah L. Chitty, then laboratory manager for the Tile Council of North America, Inc. (now the TCNA) to OSHA regarding the subject of OSHA's position regarding the COF of walkways. The letter of interpretation is posted on OSHA's website at http://www.osha.gov.

Question: Can OSHA provide any perspective or background on how the COF 0.5 value came to be attributed to OSHA?

Answer: OSHA does not have any standards that mandate a particular COF for walking/working surfaces. While there are devices to measure the COF, no OSHA standard specifically requires that employers use or have them. As you may know, there is a *non-mandatory* appendix (Appendix A to Subpart D) in the Notice of Proposed Rulemaking for Walking Working Surfaces for general industry that discusses COF. Although the notice of proposed rulemaking was published on April 10, 1990, the final rule has not yet been issued.

The pertinent portions of the *non-mandatory* appendix follow. (Note: The following appendix to Subpart D serves as a nonmandatory guideline to assist employers in complying with these sections, and to provide other helpful information. This appendix neither adds to nor detracts from the obligations contained in OSHA standards.)

Slip Resistance. A reasonable measure of slip-resistance is static coefficient of friction (COF). A COF of 0.5, which is based upon studies by the University of Michigan and reported in the "Work Surface Friction: Definitions, Laboratory and Field Measurements, and a Comprehensive Bibliography," is recommended as a guide to achieve proper slip resistance. A COF of 0.5 is not intended to be an absolute standard value. A higher COF may be necessary for certain work tasks, such as carrying objects, pushing or pulling objects, or walking up or down ramps.

Slip-resistance can vary from surface to surface, or even on the same surface, depending upon surface conditions and employee footwear. Slip-resistant flooring material such as textured, serrated, or punched surfaces and steel grating may offer additional slip-resistance. These types of floor surfaces should be installed in work areas that are generally slippery because of wet, oily, or dirty operations. Slip-resistant footwear may also be useful in reducing slipping hazards.

Although this discussion of COF appears only in a *non-mandatory* appendix in a rulemaking *proposal*, it appears to have been the basis for a statement included in an advisory appendix to the ADA Accessibility Guidelines (ADAAG) of the Access Board. The appendix statement (at A4.5) is as follows: "The Occupational Safety and Health Administration recommends that walking surfaces have a static coefficient of friction of 0.5." Together, OSHA's rulemaking proposal and the reference in ADAAG probably account for the attribution of a 0.5 COF to OSHA.

Thank you for your interest in occupational safety and health. We hope you find this information helpful. OSHA requirements are set by statute, standards, and regulations. Our interpretation letters explain these requirements and how they apply to particular circumstances, but they cannot create additional employer obligations. This letter constitutes OSHA's interpretation of the requirements discussed. Note that our enforcement guidance may be affected by changes to OSHA rules. Also, from time to time we update our guidance in response to new information. To keep apprised of such developments, you can consult OSHA's website at www.osha.gov. If you have any further questions, please feel free to contact the Office of General Industry Enforcement at (202) 693-1850.

Sincerely,
Richard E. Fairfax, Director
Directorate of Enforcement Programs

WHAT ARE THE ADA SAFETY REQUIREMENTS FOR WALKWAYS?

When the ADA was signed into law on July 26, 1990, there was a nonmandatory appendix published by the U.S. Access Board that outlined specific recommendations for walkway safety, including COF ranges for walkways and ramps, COF testing methods, and other guidelines. Because much of the information was poorly researched, confusing to understand, and often contradictory, such language was withdrawn via a technical bulletin in August 2003. A copy of the bulletin is available from the Access Board's website at http://www.access-board.gov/adaag/about/bulletins/surfaces.htm.

Since 1990, each state ratified its version of the ADA. The State of Texas adopted the ADA in 1994 under its Texas Accessibility Standards (TAS) Act.

Section 4.5 of the TAS, titled "Ground and Floor Surfaces" requirements, is as follows:

> 4.5.1 General. Ground and floor surfaces along accessible routes and in accessible rooms and spaces including floors, walks, ramps, stairs, and curb ramps, shall be stable, firm, slip-resistant, and shall comply with 4.5. *Soft or loose materials such as sand, gravel, bark, mulch or wood chips are not suitable. Cobblestone and other irregular surfaces having a texture that constitutes an obstacle or hazard, such as improperly laid flagstone, shall not be a part of accessible routes, spaces and elements.*
>
> 4.5.2 Changes in Level. Changes in level up to 1/4 in (6 mm) may be vertical and without edge treatment (see Fig. 7(c)). Changes in level between 1/4 in and 1/2 in (6 mm and 13 mm) shall be beveled with a slope no greater than 1:2 (see Fig. 7(d)). Changes in level greater than 1/2 in (13 mm) shall be accomplished by means of a ramp that complies with 4.7 or 4.8.
>
> 4.5.3 Carpet. If carpet or carpet tile is used on a ground or floor surface, then it shall be securely attached; have a firm cushion, pad, or backing, or no cushion or pad; and have a level loop, textured loop, level cut pile, or level cut/uncut pile texture. The maximum pile thickness shall be 1/2 in (13 mm) (see Fig. 8(f)). Exposed edges of carpet shall be fastened to floor surfaces and have trim along the entire length of the exposed edge. Carpet edge trim shall comply with 4.5.2.
>
> 4.5.4 Gratings. If gratings are located in walking surfaces *or along accessible routes,* then they shall have spaces no greater than 1/2 in (13 mm) wide in one direction (see Fig. 8(g)). If gratings have elongated openings, then they shall be placed so that the long dimension is perpendicular to the dominant direction of travel (see Fig. 8(h)).

Most of the ADA's guidelines have been duplicated in the ICC/ANSI A117.1 standard titled "Accessible and Usable Buildings and Facilities."

REFERENCE

Yerkes, Sara. *ICC Code Adoption Maps and Charts Codes, Standards, ANSI, OMB A-119, Governmental Consensus Process, Industry Consensus Process . . . What Is It All About?* http://www.iccsafe.org/gr/Documents/05-Who_Uses_ICC_Codes.pdf, April 23, 2003.

How and What to Measure

SLIP-RESISTANCE TEST METHODS

Numerous methods for measuring slip-resistance have been developed over the years. This section describes several of them.

The James Machine

The James Machine is a nonportable laboratory apparatus that releases a fixed weight at a preset angle onto a stationary surface to determine its dry slip resistance. Invented by UL researcher Sidney James in 1945, the James Machine is the only device recognized to measure dry SCOF per the ASTM D-2047 test method and serves as the basis for the frequently referenced 0.5 SCOF value. Although the three- by three-inch test foot of the James Machine can be modified to use any sensor material, only leather is referenced in the ASTM D-2047 and UL-410 standards.

The James Machine produces a graph describing the level of slip resistance. Interpretation of this chart is crucial. What is important when analyzing the graph is to locate the exact point where each curve begins to bend, or "break." The exact point where the curve breaks is referenced against the background scale and recorded as the SCOF.

One major drawback of the James Machine is that it is known to give erroneous data for wet surfaces. (When wet, the specification leather of ASTM D-2047 appears to act as a suction cup so that μ values are exceptionally high. They are often higher than the values obtained from the same polish when dry, which is contrary to the general pedestrian experience with wet floors.)

A HISTORICAL PROFILE OF THE JAMES MACHINE

Author Alex Sacher (1992) has written on the development of the James Machine. The following section is reproduced from his article on the topic. (Note that Sacher's original footnote numbers have been replaced by author-date citations in brackets; all references may be found in the reference section at the end of this chapter.)

In 1944, Sidney V. James of Underwriters Laboratories, Inc. (UL), presented a paper, "What Is a Safe Floor Finish?" [James 1944], at the June meeting of the Chemical Specialties Manufacturers Association (formerly National Association of Insecticide and Disinfectant Manufacturers), in which he described the development of a friction tester (the James Machine) and the correlation of test results with the safety of in-service floors. Of paramount importance in designing the instrument, following an analysis of "the mechanism of walking," was his conclusion "that the shoe is in stationary contact with the floor during the walking action" and, as such, that the slip resistance between two contacting surfaces is a function of the static coefficient of friction (SCOF).

"The basis for the judgment of the acceptability of the finish from a slipping hazard standpoint [was] that the coefficient of friction after the application of the finish shall be at least as great if not greater than that of the untreated floor surface. This is a relative or comparative method. It is based on the assumption, amply justified by experience, that the various forms of commonly used floors and floor coverings are safe enough for use without any finish material being applied.

On January 15, 1945, Sidney V. James submitted a recommendation to the Casualty Council of UL in which he informed them that

"the criterion for judgment as to acceptability for listing [e.g., of floor-covering and floor-finishing materials] has not heretofore been defined in [the usual] terms of a minimum performance specification, but in general, on a comparative basis . . . [and, further, that] a study . . . of [laboratory] test results as well as the [field] experience record covering a period of several years . . . has disclosed the fact that a minimum safe value of coefficient of friction may now be established. With our testing machine it has been found that a value of 0.50 may be set as this minimum acceptable coefficient.

Materials which have been found by experience to provide adequate underfoot safety have shown coefficients of at least 0.50. Floors and floor finishes providing appreciably less than this coefficient are found to be definitely slippery and therefore to be considered as unsafe. Abrasive-grit anti-slip treads may show values as high as 0.90 and some safe and acceptable floor waxes as low as 0.50–0.55. No material showing less than 0.50 has ever been listed by us.

It is recommended that in the future, our tests be conducted in accordance with our established standard method and that if the coefficient of friction as determined by our machine is found to be 0.50 or over, the product be recommended for listing as an acceptable anti-slip material."

This internal memorandum is the first—the seminal document, to my knowledge—in which a 0.5 SCOF value is identified as a benchmark or a watershed and equated with human locomotion safety. Significantly, it is based on a correlation of laboratory test data with extensive field experience—including feedback from several manufacturers (as noted in the James paper).

Following a number of meetings and public hearings of the floor wax and floor polish industry and government officials, the Federal Trade Commission sponsored a conference in 1948 ["F.T.C. Holds Conference" 1948] in order to resolve a number of problems concerning the 15 proposed trade practice rules. For our purposes, only one is germane. It "concerned the fact that the industry has no [agreement on a] standard method of measuring slip," that is, whether to use the Sigler (dynamic) tester or the James Machine or neither. By 1951 [*Federal Register* 1951], the Federal Trade Commission noted, "Subject to the development and acceptance of improved testing methods, either or both of the following tests with resultant coefficient of friction (COF) may be employed for the purpose of compliance with this rule:

"(1) A [dynamic] coefficient of friction of not less than 0.40 . . . Sigler test . . .
"(2) A [static] coefficient of friction of not less than 0.5 . . . by the test for slip resistance . . . by Underwriters Laboratories, Inc. (James Machine)."

At a panel discussion, "Field Testing of Waxed Floors for Slip Resistance," held by the Waxes and Floor Finishes Division of the Chemical Specialties Manufacturers Association in 1951, W. H. Joy of American Telephone & Telegraph Co., New York, stated that "whether or not a floor is safe is, to a considerable extent, a matter of opinion. We feel that our standard regular wax represents about the cutoff point [watershed] based on the background of many years [of] field experience. When tested by the Underwriters Laboratories it had a coefficient of friction on linoleum of 0.53 which is not much over their minimum of 0.50."

In 1953, the Federal Trade Commission published [*Federal Register* 1953; "Report of Committee D-21" 1951] its third (and final) set of 20 "Proposed Trade Rules for the Floor Wax and Floor Polish Industry," in which Rule 4 of the 1951 version became Rule 5—unchanged and again referencing the 0.50 SCOF as measured with the James Machine.

The American Society for Testing and Materials (ASTM) Committee D-21 on Wax Polishes and Related Materials was formed in 1950 and immediately began holding meetings on slip resistance. The D-21 Subcommittee IV on Performance Tests reported that it "is studying methods and apparatus for determining the slip resistance of waxed floors [*ASTM Bulletin* 1954]. A cooperative test program is in progress to investigate the usefulness of the Sigler Pendulum Impact machine and the Underwriters Laboratory machine for this purpose. As this work progresses, an effort will be made to correlate the laboratory results on waxed surfaces with actual slipperiness, as determined by field experience." As one might expect, the floor polish industry is uniquely positioned with respect to learning about slip-and-fall accidents.

At the 1953 annual meeting, the ASTM D-21 Committee reported that the "two proposed methods for measuring slip resistance were completed, the present plan being to publish these methods as information only. These two methods involve the use of the Sigler and James types of apparatus, respectively" ["CSMA Procedure" 1970].

Another 11 years passed before the James Machine method, alone, was accepted by the ASTM at the annual meeting in June 1964 and was issued as ASTM D-2047-64T, Tentative Method of Test for Static Coefficient of Friction of Waxed Floor Surfaces. It prescribed only the apparatus (James Machine), the test panels, the preparation of test surfaces and the leather "shoe," and the procedure—no requirement for a specific COF value. The method became a standard in 1969.

In May 1970 [Robinson and Kopf 1969], the Chemical Specialties Manufacturers Association adopted "as the Tentative Standard Test Method ASTM D-2047-69 . . . with the further provision that any such floor polish when tested by this method and has a static coefficient of friction equal to or great than 0.5 may be considered to meet the requirements of Rule 5 . . . as published by the Federal Trade Commission on March 17, 1953.

"Basically, the standard adopted by CSMA is the same as that employed by Underwriters Laboratories, Inc. . . . The only significant difference is that UL-410 requires a minimum value of 0.50 while the ASTM D-2047 standard sets a minimum of 0.5."

A paper given at the 1971 CSPA midyear meeting gave a theoretical justification for the 0.5 SCOF criteria for a safe walkway surface, based on biomechanical studies of human walking. They found that a SCOF of 0.3 should be sufficient for normal walking over a clean surface with leather-shod shoes (Ekkebus and Killey 1977).

An early report issued by UL to Hillyard Chemical Company on February 28, 1947, showed the effect of a variety of cleaning compounds and maintenance materials on asphalt tile, rubber tile, linoleum, and quarry tile. Those results showed that typical maintenance items in use at that time would maintain the surface traction level of 0.5 or higher.

The reported findings by UL indicated that slip resistance of an untreated flooring surface that may not provide a SCOF value of 0.5 could be altered through the use of maintenance products to increase the friction to the 0.5 level and beyond.

The acceptance criteria of 0.5 SCOF as measured by the James Machine offers a basis for determining if a flooring surface and maintenance system may provide the sufficient traction properties for pedestrian traffic. ASTM D-2047 remains, to date, the first and only voluntary consensus standard that specifies a compliance criterion, namely, a 0.5 SCOF.

Although a surface rendering a consistent value of 0.5 SCOF does not guarantee that accidents are not possible, it does provide for a reasonable risk for pedestrian walkways.

ASTM C-1028-06

ASTM C-1028-06 is the standard test method for determining the SCOF of ceramic tile and other like surfaces by the dynamometer pull meter method. SCOF is a term used to describe the amount of force required to cause an object (shoe outsole) to start moving across a surface (flooring material). The resultant coefficient or ratio is dimensionless to which a higher coefficient indicates increased resistance of shoe sole material to start moving across a flooring material.

Other factors can affect slip resistance, such as the degree of wear on the shoe outsole and/or flooring material; the presence of a contaminant, such as water, oil and soil; the length of the human stride at the time of slip; the type of floor finish; and the physical and mental condition of humans. Section 1.3 of the standard states,

This standard does not purport to address all of the safety concerns, if any, associated with its use. It is the responsibility of the user of this standard to establish appropriate safety and health practices and determine the applicability of regulatory limitations prior to use.

Note that the precision of this test as described in the ASTM C-1028-89 procedure indicates that the coefficient values can be expected to vary as much as 0.3. Since the COF is normally less than 1.0, test results will vary by more than 30%. The C-1028-06 version of the standard describes the precision as being "0.07 for the dry calibrated values and 0.05 for the wet calibrated values."

Also, because of the method by which ceramic tile is manufactured, the SCOF may vary within and between production runs. The SCOF of all hard-surface flooring

FIGURE 9.1
The James Machine.

materials, including ceramic tile, can be adversely affected by inadequate or improper maintenance, such as the use of unsuitable cleaning materials or procedures.

NFSI Tribometer Selection Process (TSP)

In an effort to meet its mission "to aid in the prevention of slip and fall incidents through education, training and research," it is the goal of the National Floor Safety Institute (NFSI) to effectively reduce (prevent) the ever-growing rate of slip-and-fall incidents. As a part of its research mission, the NFSI has developed a series of walkway safety standards and has encouraged manufacturers of walkway slip-resistance testing devices to produce tribometric testing equipment that meets the highest level of accuracy, reliability, portability, and ease of use.

It is the intent of the NFSI to establish walkway safety standards to meet the common goal of slip-and-fall incident prevention, the result of which impacts the manufacturing industry, insurance industry, and legal industry, all of which serve the common good of the consumer. The development of such standards will rely on the use of tribometric testing equipment, which the NFSI will certify as an effective and reliable tool for measuring walkway slip resistance. An NFSI "Approved Tribometer" (according to the April 2010 guidelines) is one that is determined by the NFSI Technical Review Committee to have met the following criteria:

1. Demonstrate laboratory accuracy of the device through submission of: Statement of Precision (per ASTM E-177-08, ASTM E-691, or equivalent) Statement of Bias (per ASTM E-177-08, ASTM E-691, or equivalent). The Statement of Precision will include published Repeatability (r) and Reproducibility (R) limits. The statements must be based on an Inter Laboratory Study (ILS) whose design fulfills the requirements covered in the *NFSI ILS Guideline*.
2. Demonstrate the field accuracy of the device through a verification process utilizing an NFSI Certified reference calibration tile.
3. The tribometer manufacturer shall be capable of providing calibration, repair, maintenance, revision control, and other services necessary to ensure device reliability.
4. The device shall be capable of measuring Static Coefficient of Friction (SCOF) to the hundredths (2 decimal places) using a scale of 0.00 to (at least) 1.00.

FIGURE 9.2
The BOT-3000.

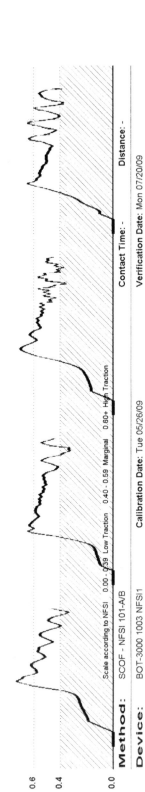

FIGURE 9.3
Sample BOT-3000 Measurement Results.

5. SCOF measurements shall be displayed via a digital display.

6. Pending the final and full release of the ASTM F13 WK6587—*Standard Practice for Validation and Calibration of Walkway Tribometers using Reference Surfaces*. The Technical Review Committee intends to add the related Validation Report and Calibration Report to this list of mandatory approval criteria.

The BOT-3000

This BOT-3000 is a robotic type of tribometer that represents the leading edge of technology in the field of slip-resistance measurement. The BOT is capable of measuring both SCOF and DCOF and is an experimental test method that the manufacturer calls "touch down friction," which simulates how a person's foot would contact the floor during normal ambulation.

The BOT-3000 is an electromechanical device whose operation and output cannot be effected by the operator, thus making it the first user-independent tribometer. And unlike tribometers which display their results on a needle gauge or scale, the BOT outputs its results on a large LCD screen as well as an optional paper graph, making data interpretation easier. The BOT also has a data port that can be used to connect the device to a laptop or desktop computer. The operator can download results in a handy PDF format for electronic filing. Rather than taking a single sample measurement, the BOT takes four individual samples and provides minimum, maximum, and average readings.

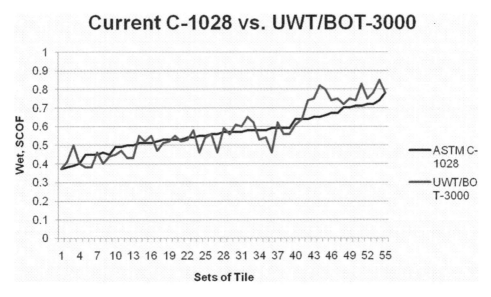

FIGURE 9.4

Surface Roughness

Another way to quantify the slip resistance of a hard-surface material is by measuring its roughness. Generally, the rougher the material, the higher its slip resistance. The smoother the material, the lower its slip resistance. Table 9.1 lists surface roughnesses for various walkway surfaces.

One of the leading studies into the effects of COF and worker safety was authored by Theodore K. Courtney and published in the *Journal of Occupational and Environmental Hygiene.* Following is an excerpt from his report (Courtney et al. 2006).

Recommended COF Values

Three ways have been used to establish the value of the friction coefficient needed for safety. The first involved anthropometry, the geometry of the walking body. By analysis of the human frame, we can see that the required friction for walking on a level surface is a function of the stride length of the walker and can be shown to be equal to the tangent of half the crotch angle of the walker. For ramps, the angle of inclination of the ramp enters the equation, and it can be described as the required friction coefficient or the tangent of the sum of half the crotch angle plus the angle of inclination of the ramp. Therefore, for equal safety, ramps should be more slip-resistant than a level floor. The second method involves kinesiology, the measurement of forces exerted by a walker's foot on the walkway surface. By the use of a force plate, the time variation of normal or contact force and tangential force that leads to cause a slip can be measured. At each instant of time, the ratio of these two forces can be calculated, giving the required coefficient of friction as a function of time. Selection of the largest value over a period of time will again result in the minimum required value of the coefficient of friction needed to prevent slipping.

The remaining method involves testing floors that have yielded a low slip-and-fall incidence rate and tracking the SCOF or DCOF values. This last method will lead to a COF "safety range" which the property owner can use as a guide in maintaining their floors and in turn serve as a management tool to prevent slips and falls.

Table 9.1.   Surface Roughnesses for Various Walkway Surfaces

Floor-Covering Material	Rz (μ)
Glazed tile	2
Porcelain	12
Ceramic GMG	17
Ceramic E	18
Natural stone	21
Granite C220	10
Concrete	60
PVC, hard	4
PVC, soft	6

It is believed that Sidney James of Underwriters Laboratories, Inc. (UL) used this third method when setting the SCOF of 0.5 for floor waxes (polishes) which in 1953 was adopted by the Federal Trade Commission (FTC) and in 1960 became ASTM D-2047. Until the 2009 release of the ANSI B101.1 standard, the ASTM D-2047 was the only standard that set a slip-resistance value against which products can be measured. Although the ASTM references the term "slip resistant" in other standards such as shoe soles, heels and various flooring materials, they have failed to provide an actual test standard by which such products can be rated.

When setting the 0.5 dry SCOF standard referenced in ASTM D-2047, Sidney James most likely used test results made with his device on floors that had given satisfactory slip performance. Later anthropometric and kinesiologic studies (Marpet 2001) have shown that for most people, wearing heels of modest height, and taking normal strides, a minimum dry SCOF value of 0.3–0.4 is required to prevent slipping. Thus the 0.5 value offers a safety factor or cushion of 1.3–1.7 (Brungraber and Templer 1992).

Methods of Assessing the Slip Resistance of Floor Surfaces

My research has shown that there are approximately 74 different devices designed to measure the slip resistance of a surface. Not all of them are designed for flooring surfaces, but of those that are, approximately 10 different methods for measuring friction have arisen. In every one of these, the test parameters can be varied so that the ranking of the different material combinations tested can be changed, producing hundreds of method variants. Standberg (1983) estimated that of the more than 70 slip testers that have been developed to determine the COF of a surface, most have been designed as modified versions of either the drag sled, an inclined plane, an inclined leg, a swinging pendulum, a slider (test foot) apparatus, or an electronic test rig incorporating a force platform. It should be recognized that the COFs obtained with these various methods and machines do not necessarily correlate empirically and because of such have led to a great deal of controversy as to which device is the best.

ISO Technical Committee 189 (ceramic tiles) has spent a great deal of time reviewing the various test methods and has paid particular attention to the pendulum test, the Tortus test (a slider apparatus), and the ramp test (an inclined plane), each of which measures DCOF. After extensive deliberation and some round-robin testing, ISO/TC 189 Working Group I is considering recommending the adoption of a slider-type apparatus (the Tortus) for nonprofiled ceramic tiles. Heavily profiled tiles would be tested using the German ramp method ("Ideas for OSHA" 2006).

The ramp test is considered most suited to assessing the slip resistance of industrial slip-resistant tiles, which are coarsely profiled and provide so-called displacement spaces between studs, blocks, ripples, and the like. Such tiles impart slip resistance because of the nature of the surface texture rather than the COF of the material. Because of the deep profile of these tiles, the test foot can make only limited contact with the tile and skim the surface, thus not providing a true and accurate depiction of the tiles' slip-resistance performance.

The question of how best to bridge the gap between European and American interests took a turn for the better in 2009 with Jens Sebald's publication of *System Oriented Concept for Testing and Assessment of the Slip Resistance of Safety, Protective and Occupational Footwear*. Sebald's research highlighted the linear relationship between the (German) ramp, the tribometer of choice for most of Europe, and the BOT-3000, the state-of-the-art robotic device that has gained favor in the United States. Because both the ramp and the BOT-3000 produce very similar results, the opportunity for European ramp users to now take measurements in the field has become a reality. The BOT-3000 is a portable ramp tester. Sebald also describes what he calls the "Friction System," or relationship between the four "Friction Partners," which include the flooring surface, the individual's footwear, the contaminants that separate the shoe–floor interface, and the ambient effects of temperature and humidity.

Studies of the magnitude and duration of actual forces produced in the shoe–floor interface during normal strides have revealed that nearly all subjects, regardless of footwear, produce characteristic instantaneous horizontal and vertical force peaks that correspond almost exactly to the same phase of leg and foot position from person to person. The chance of a dangerous/uncontrollable slip is the greatest at two points; first, as the heel tip first contacts the floor before the foot is planted firmly and while it is still moving forward, and, second, momentarily after the foot has stopped and while the subject's weight is shifting onto the heel during forward rocking motion (Kohr 1992).

Data from such subjects and data related to various types of footwear and flooring surfaces need to be considered when designing and assessing slip-resistance testing apparatuses. Crucial to developing a meaningful method is recognizing that the measured slip resistance between two surfaces is specific to that pair of materials under those test conditions, thus the importance of the Friction Partners. A single material does not have one unique slip-resistance characteristic, and therefore it is misleading to suggest that slip performance between one set of Friction Partners is the same as that of a different pair of Friction Partners (Carani et al. 1992).

SCOF versus DCOF

In the United States, the standard method of measuring slip resistance is that of SCOF. However, it has been increasingly recognized ("Working Surfaces/Slips and Falls" 2000) that falls result from a movement of the contacting shoe across a surface. Therefore, the lower the DCOF, the less likely the individual is to catch him- or herself and prevent falling. The greater the DCOF, the more likely a slip can be controlled. While a standard DCOF test method has not been developed in the United States, it seems to be generally agreed that DCOF may provide the best correlation as it relates to the real-world walking-and-slipping experience (Barnett 2002; Bonnett et al. 2004; Courtney et al. 2006; Kohr 1992).

A dry SCOF of 0.5 is considered by many as a threshold for safety of surface, without considering that this threshold is related to the test method and the conditions that it models. The 0.5 SCOF value has served as the minimum safe level of static friction and has carried pseudolegal status in the United States (Sebald 2009). The 0.5 value has been supported by both a theoretical analysis as well as experiential history. In reality, this single numeric value has proven to be more complex, as it does not take into account the friction-velocity characteristic of the two contacting surfaces.

Investigators have found that in walking, a person's foot is not moving parallel to the floor except during a slip (Courtney et al. 2006). Thus, in order to slip, you must first start to feel yourself slipping before you can overcome the SCOF. Once a slip has started, it depends on the DCOF, which is usually less than SCOF and is a function of a person's body geometry (anthropometry). For a slip to be controlled or stopped, there must be a rapid and significant change of the body geometry, which is often difficult to accomplish, especially if you are elderly. As a result, slips, once started, are often uncontrollable and usually result in a fall.

When Barrett's criteria as seen in table 9.2 were first published, the importance of kinetic (dynamic) friction was recognized, and the possibility of dynamic friction being greater than static friction was first presented. It has now been recognized that the SCOF of 0.5 is, at best, a guideline for normal level-surface ambulation and does not represent either the maximum or the minimum required COF required for all slip-resistant pedestrian traction requirements ("Working Surfaces/Slips and Falls" 2000).

It has been claimed that a DCOF value of 0.4 is a reasonable safety guideline for a normal-length stride, and the relationship between step length and the COF necessary for stability has been calculated (Sebald 2009). No matter how well the 0.4 value correlates to other experimentally determined values, it cannot be regarded as a blanket threshold of safety. Because of the variables between the various test methods, one cannot simply suggest that a floor is safe or unsafe unless all the variables have been accounted for.

Table 9.2.   Barrett's Criteria for Surfaces Based on the COF

SCOF	DCOF	Classification
Above 1.1	Greater than static	Safe and rising
	Equal to static	Safe
	Less than static	Safe but falling
0.5–1.1	Greater than static	Potentially safe
	Equal to static	Intermediate
Below 0.5	Greater than static	Slippery but rising
	Equal to static	Slippery
	Less than static	Slippery and falling

The concept of slip resistance is complex and related to many factors. In dry conditions, the COF is the most important factor, while in wet conditions, surface roughness is equally important since it provides for the migration of water from beneath the sole–floor interface, thus increasing the contact area and, in turn, the COF.

Which Tribometer Is the Best?

In an ideal world, all slip testers of the same type should deliver identical results; however, in the real world, that is not necessarily the case. The measurement of walkway friction is inherently plagued by the fact that the slip-and-fall equation has more variables than constants and is therefore not easily solved. Walkway surfaces often vary substantially from point to point and over time. The test-foot material and test-surface materials may change with the accumulation of contaminants as well as time. And different tribometer types can give substantially different results because of a non–Amontons-Coulomb friction relationship. Surface friction is not only a property of a material but a system property as well.

All ASTM standard test methods require a statement of the testing apparatuses' precision and bias. The precision of a method concerns its reliability and reproducibility. Reliability is the ability of a given instrument/operator to obtain consistent results. Reproducibility is the ability of different operators/laboratories to obtain similar results. The *bias* of a particular test instrument addresses the ability of the instrument to correctly identify the correct value of a specific parameter. For example, a butcher's weight scale would be biased as giving consistently high results if the operator of the scale consistently placed his thumb on the weighing tray.

For a number of reasons, the establishment of slip-resistance standards has become highly adversarial and chaotic. The ASTM F-13 Committee on "Footwear and Traction" has been at the center of this firestorm. Manufacturers fear that if a standard is introduced that will allow for the use of a portable/nonlaboratory testing device, such an instrument will become a weapon to be used against them in court and by regulatory agencies. Others take the position that property owners have the right to know if the floors in their business are in fact slip resistant and safe for their employees and invited guests.

As a result of the all-too-frequent inter- and intracommittee turf battles in which the ASTM F-13 Committee seemed to be involved, it soon found itself a target of the ASTM Committee on Standards (COS). As a result, COS issued a directive that F-13 develop standards in a way that would minimize "marketplace confusion" and set F-13 on a timetable to achieve approval of the existing standard's Precision and Bias Statement.

The reason that F-13 has found itself in trouble with respect to Precision and Bias is that the COS had set hard deadlines as to when the Precision and Bias Statements had to be incorporated into the existing standards that neither the F-1677 nor the F-1679 standards were able to meet. To add insult to injury, associated with each standard was a propriety device, both of which carried U.S. patents, another violation of ASTM operating procedures. In the end, the ASTM did what had to be done and withdrew both standards. Just when you thought the debate over F-1677 and F-1679 was over and that reasonable, fair-minded people have entered the leadership of the committee, they proposed a new standard, almost identical in scope to that of the VIT standard, this time renaming the device the VAT, or variable angle tribometer.

The Key Is in Cleaning the Sensor Material

Tribometric test feet have in the past been surfaced with standard leather (D-2047), Neolite® Test Liner ASTM C-1028, NFSI 101-A/B, and ANSI B101.1. European countries have chosen to use Rapra 4S (4S rubber) as their reference material. Each of these materials is far from ideal:

- Standard leather does not come from standard cows. It is subject to the variability inherent in animal skin. Living in the twenty-first century, it seems a bit archaic to use dried animal flesh as a means of scientific measurement, but it is what it is.
- Neolite® Test Liner is not a shoe-bottom material. It had originally been developed to test cobbler's adhesives. Questions have been raised about the quality control that goes into the production of Neolite Test Liner as well as the single source for the material. Even more alarming are the recent announcements by Smithers Laboratories, the exclusive manufacturer of laboratory-grade Neolite, that it may soon discontinue production.
- Rapra products are not readily available in the United States, and they are quite costly. Surprisingly, 4S rubber is quite abundant and inexpensive in Europe.

Many researchers as well as the footwear industry have promoted the use of real footwear outsole compounds, such as Nitrile rubber, to be used as tribometric test-foot reference materials. Because such materials reflect real-world performance, many believe that the use of such material will make tribometric test results inherently more credible.

In any case, the tribometric reference-material sets will form the bedrock of all tribometer verification and friction testing. *Ideally*, the composition of the reference-pair sets will consist of the following:

1. A single test surface on which a suite of different test-foot materials will be tested. Again, the different test-foot materials, which will range from *very* slippery to *very* tractive, will all have real-world shoe-bottom applications.
2. A single test-foot material under which a suite of different test surfaces, ranging between *very* slippery and *very* tractive, will be tested. Again, *ideally*, each of the tribometric test-surface reference materials will have real-world application as a floor

surface (this does seem unrealistic, especially on the low-friction end of the scale, as floors are rarely if ever designed explicitly to be truly slippery) [Marpet 2001].

Measurement is always first made of the floor surface after brushing only so that the effect of any contamination can be known [Brough et al. 1979]. Although often used interchangeably, terms such as "nonslip," "nonskid," "skid resistant," "slip retardant," and "antislip" remain undefined and therefore misleading. Their continued use, mainly by product manufacturers, is used primarily as a result of minimizing liability. After all, you cannot sue a product manufacturer for claims made on a product whose slip characteristics are undefined. The only terms that are defined by way of a scientific test method are "slip resistant" and "high traction." A slip-resistant surface is one whose dry SCOF is 0.5 or greater when measured via the James Machine. The definition is restricted to floor polishes and finishes. High-traction surfaces are those that possess a wet SCOF of greater than 0.5 per the NFSI 101-A/B standards or ANSI B101.1. With a few exceptions, this standard applies to most hard-surface walkways.

"Slip Resistant" versus "Slip Resistance"

Tisserand (1985) describes slip resistance this way:

> Slip resistance is a function of many factors, among which the coefficient of friction is only one, albeit probably the most important. At best, it is a descriptive term, encompassing all the critical material and human elements which may lead to a slip, and as such should not be used interchangeably with coefficient of friction. Slip resistance is neither a constant nor an intrinsic property of a given surface, be it flooring, floor coating, or footwear. Instead, it is an ephemeral characteristic, determined by the activities at the time, whether one is walking naturally, walking fast or running, turning sharply, pulling or pushing a load, or going up or down an inclined plane or steps, coupled with the physiological, perceptual, and behavioral condition of the individual.

Moreover, slip resistance is a function of surface texture, wear, and cleanliness of both the walking surface and the footwear outsole.

Table 9.3.   DCOF for Leather Heels

Surface	Wet	Dry
Ice		.08–.10
Terrazzo, marble	.17	.33
Linoleum	.11	.33
Ceramic tile	.19	.36
White oak (waxed)	.17	.24
Concrete	.37	.43
Steel plate	.19	.49
Treated steel plate	.63	.70

Source: Ceramic Engineering and Science Proceedings.

Table 9.4. SCOF for Leather and Neoprene Heels

Surface	Leather	Neoprene
VCT, dry	.46	.58
VCT, wet	.30	.63
Linoleum, dry	.27	.42
Terrazzo, dry	.25	.38

Source: Ceramic Engineering and Science Proceedings.

- Glazed non-antislip tiles have the least favorable behavior since, while they have almost the same values of the coefficient of friction in dry conditions as other kinds of tile, the values of the coefficient of friction decrease considerably in wet conditions, reaching values of dangerous slipperiness.
- Unglazed non-antislip tiles, glazed antislip tiles, and unglazed antislip tiles with a rough texture are very similar, in particular with regard to the median values.
- Unglazed antislip tiles with surface relief are the best products in regard to slip resistance.

In any case, for all types of products, the range of COF values is relatively wide. This means that in each group, products can be found that have widely different performances in regard to their antislip characteristics.

Surface textures greater than 2 mm are very effective in breaking the liquid film layer separating the floor from the person's footwear, thus providing better traction. This depth of relief is directly related to an increase in slip resistance as well as the cleanability and durability of the tile surface. Simply put, the rougher the flooring surface, the harder it is to keep clean.

Table 9.5. Comparison of Foot Materials on Tile

Outsole Material	Shore A Hardness	COF at: 0.3 cm s–1 Dry	Wet	7.4 cm s–1 Wet
Best-grade leather	95	0.67	0.76	0.72
TRRL skid-tester rubber	55	1.67	0.95	1.12
Crepe rubber	38	1.06	0.76	0.90
Solid gristle rubber	67	0.93	0.79	1.00
PVC	78	0.78	0.79	1.00
Composition rubber	84	0.82	0.88	0.91
SBR microcellular rubber	51	1.07	0.92	1.09
Reaction molded polyurethane	65	0.90	0.92	1.07
EVA microplastic rubber	54	0.60	0.61	0.66
Thermoplastic rubber	36	1.07	0.81	1.01
Hard-rubber composition	90	0.89	0.86	0.91
Resin rubber	95	0.72	0.77	0.83

Source: Brough et al. (1979).

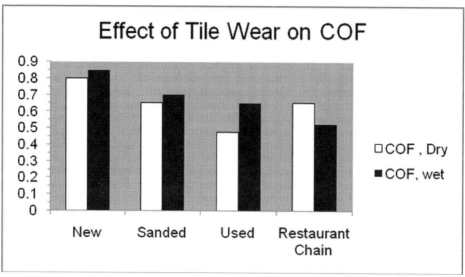

FIGURE 9.5

Research has shown that the best performance from the point of view of slip resistance, but especially in regard to maintaining a high level of slip resistance in wet conditions (i.e., when the floor is covered with a thin film of water), was found for unglazed tile with surface relief having a depth of relief of the order of 2 to 3 mm (Carani et al. 1992).

WALKWAY AUDITING

Third-Party Consultation Program

Several years ago, the American Society of Safety Engineers (ASSE) supported a bill introduced by U.S. Senator Michael Enzi that would allow third-party safety audits of companies. It was claimed that "such audits, conducted under strict requirements to ensure professionalism, would serve as yet another opportunity for small and medium-sized employers to improve workplace safety and health. We would not support such a program if we believed it would jeopardize the capabilities of existing federal or state consultation programs. Too many employers do not seek assistance at all, so we must attempt to find new ways" ("Ideas for OSHA" 2006).

Although simple, the idea that a property owner could actually test the slip resistance of a floor so as to identify the risk of a slip and fall did not occur until the NFSI authored the first comprehensive standard for safety. The ANSI/NFSI B101.1-2009 standard, coupled with a comprehensive walkway auditor training program, has already yielded positive results in slip-and-fall prevention.

The purpose of conducting a floor audit is to first identify areas of a walkway that may present an elevated risk of a slip and fall and second to provide the property owner recommendations by which the situation can be remediated. The floor audit should not be used as a litigation tool whereby COF readings taken in the present are used to suggest a level of safety in the past.

Another important aspect of walkway auditing is that it allows property owners the opportunity to track their floors' COF performance over time. One example is that of a grocery store chain that will periodically test its floors' wet SCOF values to see if they are in compliance with both industry standards and their in-house slip-resistance standard. If a customer was to accidentally spill a product onto the floor, which is common in the grocery industry, that the store did not properly clean, the audit would reveal the slippery condition. Without the periodic audit, the hazardous condition would go unknown and could reveal itself by way of a slip-and-fall accident.

How can you claim that your floors were "slip resistant" when you have not measured their COF? Just because the floor was manufactured with a high SCOF does not mean that it will maintain such a level of safety. The property owner should demonstrate that floors are properly cleaned and provide frequent inspections. The addition of audit data will serve as the basis for claiming that floors were indeed slip resistant.

This concept is also true for the household consumer. Several companies provide a "do-it-yourself" floor audit whereby, for a small fee, home owners can rent a portable slip tester to measure their floors' wet SCOF. Although such testers are not as accurate as those costing thousands of dollars more, they can provide a relatively accurate picture of a floor's level of slip resistance, and the home owner can then determine what if any action to take to correct the problem. In most cases, the reason that residential floors become slippery is improper maintenance. Often a residential floor can be corrected simply by changing the cleaning product. Products that contain high levels of detergents and fragrances can leave behind a slippery residue.

The recommendation of the Expert Panel on Reducing the Number of Slips and Falls in Canadian School District Facilities to the Canadian school districts was simple: purchase a slip-test meter. The panel's 2004 study found that standardizing the use of a slip meter would provide the following benefits:

- Allow for exchange of measurements with adjoining school districts
- Standardize on a measurement technique compatible with other school districts

- Monitor the after-application wear history of flooring regularly and record the results
- Use the tribometer readings as a tool to establish re-treatment intervals based on COF readings
- Build a database on flooring performance over time and floor treatment performance over time
- Exchange slip-meter measurement information and floor treatment performance with other school districts
- Use the tribometer as part of the incident investigation process (to assess "slipperiness")

Their reasoning? Slips and falls happen when floor finish erodes and when the floor is not kept clean. School districts allocate considerable resources to prevent erosion and to ensure cleanliness. School district maintenance groups function in isolation and often are unaware of what other districts are doing to prevent slips and falls. If one district identifies a superior product (or an inferior one), there is no defined mechanism by which individual schools can distribute information. There is also nothing other than anecdotal information available to convince another district of the merits of their discovery. "On those rare situations where the contaminant is a failed product and is not visible, routine testing of floor conditions with an inexpensive slip meter is useful. It is useful primarily as an aid to intuition and plays only a small part in a systematic program. . . . The trick is to isolate the contaminants and remove them, not to measure them. Your most important asset is your employee, not your slip meter. If School Districts could agree on a common slip-test device and, if Districts would commit to using the device to evaluate the performance of floor materials and floor treatment products, Districts with the same environmental challenges would be able to select products with a proven performance history. Products that had failed to realize their potential in one District would be flagged and another District considering the use would be forewarned." The committee's executive summary is reproduced in the appendix to this chapter. An example of a walkway audit modified from Traction Auditing, LLC, follows.

SIX STEPS TO PREVENTING SLIPS AND FALLS

1. **Select the Right Floor.** Today's consumers have hundreds of choices when selecting floors for their homes. However, most consumers make their decision based on a combination of color, pattern, and cost and may later find that the floor they chose is more slippery than they realized. Also, materials such as linoleum and vinyl,

Example of a Walkway Audit

FIELD TEST REPORT

TEST DATE	08/01/2009
TEST NUMBER	001
TIME IN	0955
TIME OUT	1055
REPORT DATE	08/05/2009
CLIENT NUMBER	1234
CASE NUMBER	NA
CLIENT NAME	Catfish Joe's
CLIENT ADDRESS	101 Main Street
CITY, STATE, ZIP	Islip, NY 11751
PHONE/FAX	_____
TEST PERFORMED AT	Catfish Joe's Restaurant
CONTACT PERSON	Catfish Joe
PHONE/EMAIL	_____
CONTACT SUPERVISOR	N/A
PHONE/EMAIL	_____
HOURS OF OPERATION	1100- 2000, Thurs-Sat
INSURANCE CARRIER:	
GL	N/A
WORKERS COMP	N/A
INSURANCE CONTACT	N/A
MAINTENANCE CONTRACTOR	N/A
CLEANING PRODUCTS USED	Neutral floor cleaner/degreaser
CLEANING PROTOCOL AVAILABLE	Yes ☒ No ☐
MAINTENANCE LOG AVAILABLE	Yes ☐ No ☒
PREMISES INSPECTION LOG	
AVAILABLE (WALK AROUND)	Yes ☐ No ☒
TIME INTERVAL	N/A
AUDITOR NAME	Sam Safe
CERTIFICATION NUMBER(S)	NFSI #0001
TRIBOMETER TYPE	BOT-3000
TRIBOMETER ID #	1000
DATE CALIBRATED	6-24-08
DATE VERIFIED	8-1-09

OBSERVATIONS

FLOORING TYPE
12" × 12" ceramic tile in dining room
4" × 4" quarry tile in kitchen

FLOORING MANUFACTURER Unknown

DATES FLOORING INSTALLED Original to occupant — 7 years

SURFACE CONDITION
☐ Clean ☐ Contaminated ☒ Smooth ☒ Rough ☐ Even ☒ Uneven ☐
Change in Elevation
☐ Other (Explain) N/A

SURFACE TEXTURE
☒ Smooth ☐ Grooved ☐ Course ☐ Rough ☐ Patterned ☐ Abrasive Added
☐ Other (Explain) N/A

STRUCTURAL IRREGULARITIES None

SURFACE PREPARATION   ☐ Sealed ☒ Unsealed

TYPE OF SEALER   N/A

COMMENT   *None*

SURFACE CARE
☐ Dust mop ☐ Dry Mop ☐ Wet Mop ☒ Deck Brush/Mop ☐ Deck
Brush/Vac ☐ Buff/Burnish ☐ Floor Machine
Brand _____ Type _____ Model _____

CONTAMINANTS OBSERVED
Grout line and texture holding soil/contaminants

SAFETY PROGRAM IN PLACE   ☐ Yes ☒ No ☐ N/A

CAUTION SIGNS AVAILABLE   ☐ Yes ☐ No ☒ N/A

COMMENTS

Walkways appeared clean, however upon closer inspection both the grout lines and tile texture appeared soiled/contaminated to which I confirmed upon touch. According to the owner, the floor is cleaned each morning prior to opening the restaurant. The owner stated that the mop head is changed approximately once a week and the mop water/solution is changed daily. The floors are mopped via the use of a neutral cleaner/degreaser purchased at Sam's Club. After mopping the floors are rinsed and allowed to air dry. No written maintenance policies exist for maintenance. Restaurant employees are verbally trained as to the maintenance procedures.

SITE DIAGRAM:

[Auditor draws site diagram here.]

TEST DATA

BUILDING NAME   Catfish Joe's Restaurant
ROOM NUMBER(S)   Dining Room, Entrance, and Kitchen
ZONE NUMBER   1   **RISK CLASS**   B

EXAMPLE DATA READINGS: CATFISH JOE'S RESTAURANT

Location Number	Test Type, Wet/Dry	Sensor Type, Leather/Neo	COF, Test 1	COF, Test 2	COF, Average
01	Wet	Neo	0.59	0.60	0.60
02	Wet	Neo	0.39	0.40	0.40
03	Wet	Neo	0.48	0.47	0.48
04	Wet	Neo	0.58	0.50	0.54
05	Wet	Neo	0.49	0.48	0.49
06	Wet	Neo	0.45	0.43	0.44
07	Wet	Neo	0.31	0.35	0.33

NOTE: Shaded measurements are deficient to that of the NFSI 101-A standard.

DATA INTERPRETATION: Data readings were taken per the NFSI 101-A standard. A system of three basic SCOF ranges has been charted below.

NFSI 101-A TRACTION RANGES

Wet SCOF Value (μ)	Available Traction	Remediation
$\mu \geq 0.60$	High traction: lower probability of slipping	Monitor SCOF regularly and maintain cleanliness
$0.40 \leq \mu < 0.60$	Moderate traction: increased probability of slipping	Monitor SCOF regularly and maintain cleanliness. Consider traction-enhancing products and technologies
$\mu < 0.40$	Minimal available traction: higher probability of slipping	Seek professional intervention. Consider replacing flooring and/or coating with high-traction products

These categories are not indicative of all possible conditions. There are numerous variables that may add to or take from the available traction of any given floor surface (type or style of footwear, type and frequency of contaminants, pedestrian preoccupation, etc.). Results of dry and wet tests should be viewed independent of each other and not compared.

CONCLUSION

The grouted tiled walkways at Catfish Joe's Restaurant were tested for their wet Static Coefficient of Friction (SCOF) per the NFSI 101-A standard using a BOT-3000 tribometer. The BOT-3000 is an approved testing device per the NFSI. Seven locations labeled 01-07, were selected as representative samples of the walkway environment. Each location was tested twice in two directions each of which were perpendicular to each other. The NFSI 101-A standard defines a "High-Traction" surface as that which exhibits a wet SCOF of 0.6.

Measurement 01 was taken beneath the carpet mats, and measurements 05 and 06 were taken adjacent to the carpet mats. The section of tile beneath the carpet mat had an average wet SCOF value of 0.6 thus indicating that it was a High-Traction surface. Locations 02, 03, 04, 05, and 06 each had an average wet SCOF value between 0.4 and 0.5, thus placing them in the Moderate Traction range. Finally, Location 07 was that of the kitchen floor which had an average wet SCOF of 0.33 placing it in the Minimal Available Traction range and thus presented the highest level of risk in the restaurant.

Based on the data listed above it is my recommendation that the restaurant consider using an NFSI-Certified floor cleaning product to enhance the slip resistance of the dining room floor. Although it appears that the floor mats were working to improve the safety of the customers near the buffet line, floor mats provide only an "island of safety" and may present an elevated trip hazard.

Finally, it is imperative that immediate action be taken to improve the slip resistance of the kitchen floor. On average, quarry tile like that used in the restaurant kitchen has a wet SCOF of 0.8 or greater when new. Based on my measurements, the floor in question has seen more than a 60% decrease in slip resistance, most likely caused by improper cleaning. The use of an NFSI safety treatment and/or daily maintenance product is recommended to enhance the safety of the kitchen floor.

which wear over time, may result in a decrease in slip resistance. Similarly, materials like polished marble or granite are very attractive but offer poor slip resistance, especially when wet, and therefore should not be used in high-risk areas such as bathrooms or kitchens. Ironically, such locations are exactly where most designers and home owners will use such products, only to later realize the increased slip risk they have created for themselves and their families. At last, a new labeling system is being developed to make it easier for consumers to know just how slippery a flooring product is before they purchase it. The soon-to-be-released ANSI B101.5 labeling standard will provide a simple, easy-to-read label identifying the traction level of new floors. Flooring materials will be labeled as either high traction, moderate traction, or low traction in order to assist the consumer in making a more educated buying decision as it relates to slip resistance.

2. **Choose the Right Floor Cleaner.** The NFSI's study on household floor cleaners showed that 10 out of 19 household floor cleaners actually made floors more slippery than before use. If improperly used, any floor cleaner can create a slippery film. One example is that of wood floors. Many of today's wood floors are actually laminated or engineered and have a permanent wear layer applied at the factory. And although the floor is made of wood, the wear layer is actually plastic. Such floors should be maintained with a product designed to clean a synthetic surface rather than a wood-floor cleaner. Wood-floor cleaners contain oils designed to penetrate into the wood's naturally porous surface to maintain the wood's moisture level and to prevent cracking. The use of wood-floor cleaners on a nonporous floor will often result in a slippery/oily film, which may lead to a slip and fall.

3. **Wear Appropriate Footwear.** The NFSI estimates that 24% of all slip-and-fall claims are directly caused by inappropriate footwear. Proper selection of footwear can make a big difference in preventing slips and falls. Workers should chose rubber-soled shoes that are labeled as "slip resistant." Slip-resistant soles offer the highest level of slip resistance when walking on wet or oily floors. In addition, employees should avoid wearing shoes that provide poor slip resistance, such as flip-flops. Loose-fitting, foam-soled shoes such as flip-flops offer very poor slip resistance under wet conditions and have been linked to a growing number of slip, trip, and fall claims. Finally, it has been estimated that as many as 60% of Americans wear shoes that are not their proper size. Women tend to wear shoes that are smaller than their actual foot size, while men tend to wear larger, wider shoes than needed. Either way, wearing the wrong-size shoe can change the way you walk and, in turn, lead to a slip and fall or a trip and fall.

4. **Remove Walkway Hazards.** Is your house an accident ready to happen? Take a minute to audit your home and look for hazards such as electrical cords, frayed carpet edges, and loose stair edges. Many hazards actually become common for

most home owners who simply get used to that loose stairway handrail or buckled carpet, often resulting in a fall at some point. Don't be complacent when it comes to safety. Just because you have been accustomed to household hazards does not mean that they no longer present a danger—they do. The top 10 household slip, trip, and fall hazards include the following:

i.   Buckled or frayed carpeting
ii.   Loose or buckled kitchen throw rugs
iii.   Exposed electrical cords
iv.   Loose or missing stairway handrails
v.   Loose, worn, or missing stair edges
vi.   Slippery or excessively worn floors
vii.   Cracked or elevated floor tiles
viii. Broken, cracked, or icy sidewalks
ix.   Clutter and debris
x.   Slippery bathtubs

5. **Improve Lighting.** If you can't see a hazardous condition, you are more likely to get hurt. Improving the lighting conditions in your home can play a big role in identifying walkway hazards. Focus on the "three ways": entranceways, stairways, and hallways. As you enter a home through an exterior entrance, your eyes need time to adjust to a change in lighting. This is also true for stairways and interior hallways, which are often dimly lighted. Increased lighting in these three locations can play a significant role in fall prevention. Also, remember that as an incandescent lightbulb ages, the light it omits decreases. White light is better than yellow light. Incandescent lights produce light shifted toward the yellow range of the spectrum, while halogen or LED lighting produces a whiter, brighter range of light. Use of halogen or LED lighting can often increase color contrast and make potential walkway hazards more visible.

6. **Exercise and Improve Balance.** It's up to you to protect yourself from being the victim of a fall, and like lightbulbs, whose glow decreases over time, the same is true for us. As we age, our muscles and bones change and can lead to a loss of balance. Stay in shape. Remain active. Increased muscle tone will lead to an increase in balance, mobility, and reaction time. Finally, research has shown that as we walk, we only raise our feet approximately a quarter of an inch above the ground we are walking on. As we age, our gait changes too as we begin to take shorter, choppier steps and eventually begin to shuffle rather than walk, making even the smallest walkway elevations a potential trip-and-fall hazard. However, through exercise and improved balance, most people can reduce their risk of a fall.

REFERENCES

ASTM Bulletin, January 1954, p. 16.

Barnett, Ralph Lipsey. "'Slip and Fall' Theory—Extreme Order Statistics." *International Journal of Occupational Safety and Ergonomics (JOSE)* 8 (2, 2002): 135–158.

Bonnet, John, et al. Expert Panel on Reducing the Number of Slips/Falls in School District Facilities, August 2004, BCSSA/BCPSEA/WCB.

Brough, R., F. Malkin, and R. Harrison. "Measurement of the Coefficient of Friction of Floors." *Journal of Physics D: Applied Physics* 12 (1979): 517.

Brungraber Robert, and John Templer. "Controlled Slip Resistance." *Ceramic Engineering and Science Proceedings* 13, no. 1–2 (1992): 66–77.

Carani, G., A. Tenaglia, and G. Timellini. "Slip Resistance of Ceramic Floor Tile: Design Criteria for Anti-Slip Tile." *Ceramic Engineering and Science Proceedings* 13 (1–2, 1992): 1–13.

Courtney, Theodore K., Yeung-Hsiang Huang, Santosh K. Verma, Wen-Ruey Chang, Kai Way Li, and Alfred J. Filiaggi. "Factors Influencing Restaurant Worker Perception of Floor Slipperiness." *Journal of Occupational and Environmental Hygiene* 3 (2006): 593–599.

"CSMA Procedure for Products Classified as to Slip Resistance," Chemical Specialties Manufacturers Association, *CSMA Bulletin 308-70*, December 8, 1970.

Ekkebus, C. F., and W. Killey. "Validity of 0.5 Static Coefficient of Friction (James Machine) as a Measure of Safe Walkway Surfaces." Test Methods and General Information, January 1977, 277–280. Chemical Specialties Manufacturers Association, Inc.

Federal Register, Tuesday, April 17, 1951, p. 3360.

Federal Register, Tuesday, March 17, 1953, p. 1513.

"Field Testing of Waxed Floors for Slip Resistance," *Soap and Sanitary Chemicals* (October 1951): 138–142, 151, 153.

"F.T.C. Holds Conference on Floor Wax Trace Practices," *Soap and Sanitary Chemicals* (September 1948): 145, 149.

Harris, G. W., and S. R. Shaw. *Journal of Occupational Accidents* 9 (1988): 287–298.

"Ideas for OSHA." *Professional Safety*, August 2006, 17.

James, S. V. "What Is a Safe Floor Finish?," *Soap and Sanitary Chemicals* 20 (October 1944): 111–115.

Keyersling, W. Monroe. *IOE 539 Notes: Working Surfaces/Slips and Falls.* http://ioe.engin.umich.edu/ioe539/slips.pdf, July 24, 2000.

Kohr, Robert L. "A Comparative Analysis of the Slipperiness of Floor Cleaning Chemicals Using Three Slip Measuring Devices." *Ceramic Engineering and Science Proceedings* 13 (1–2, 1992): 14–21.

Marpet, Mark I. "Problems and Progress in the Development of Standards for Quantifying Friction at the Walkway Interface." *Tribology International* 34 (2001): 635–645.

"Report of Committee D-21 on Wax Polishes and Related Materials," pp. 434–435 in *ASTM Proceedings*, Vol. 51, 1951.

Robinson, W. H., and R. E. Kopf, "Evaluation of the Horizontal Pull Slipmeter," *Materials Research & Standards* (July 1969): 22–24.

Sacher, Alex. "Is the 0.5 Static Coefficient of Friction Value a Bench Mark or a Watershed?" *Ceramic Engineering and Science Proceedings* 13 (1–2, 1992): 29–45.

Sebald, Jens. *System Oriented Concept for Testing and Assessment of the Slip Resistance of Safety, Protective and Occupational Footwear.* © 2009 by Pro BUSINESS GmbH, ISBN 978-3-86805-356-2. An English translation is available from the National Floor Safety Institute at http://www.nfsi.org.

Standberg, L. "Ergonomics Applied to Slipping Accidents." In *Ergonomics of Work Station Design*, edited by T. Kvalseth. Stoneham, MA: Butterworth, 1983, 201–228.

Tisserand, M. "Progress in the Prevention of Falls Caused by Slipping," *Ergonomics* 28(7) (1985): 1027-1042.

Appendix

EXPERT PANEL ON REDUCING THE NUMBER OF SLIPS/FALLS IN SCHOOL DISTRICT FACILITIES PUBLISHED: 2004, EXECUTIVE SUMMARY
Note: This appendix is reproduced from the Expert Panel's final report.

In April 2004, an Expert Panel was selected and charged with investigating the cause of slip/fall incidents in School District facilities. Over the following eight-week period, the Expert Panel focused its attention on slips/falls that occur indoors and on the same level.

The Expert Panel began its work by attempting to collect data from the WCB and from contributing School Districts. The Panel discovered that data is available, however, it has not been acquired in a manner that would allow outside reviewers to focus on the root cause of incidents and the contributing factors. The data did, however, confirm that slips and falls did occur and in the same general frequency as appears in the literature (that 15–20% of total claims are derived from slips/falls). On that basis, it was evident that slips and falls cost School Districts a considerable sum of money (in staff replacement costs, in time-loss absences, in medical costs, in rehabilitation costs and in pensions). The panel concluded that reviewing slip/fall incidents from a prevention perspective was warranted.

The Panel canvassed Occupational Health and Safety Agencies across Canada. None of the Agencies had studied the slip/fall issue from this perspective. The Agencies were interested but were unable to provide information.

The Expert Panel quickly determined that the slip/fall problem would not be easy to resolve. Many factors contribute to slips and falls. Only a few factors are actually under the immediate control of a building owner/operator.

The Expert Panel determined that selection of flooring material was a significant factor in slip/fall incidents and was one that would be within the control of the building owner/operator. The Panel found that looks alone should not dictate the choice of flooring material; choice required careful consideration of durability, use pattern, environmental factors and maintenance capabilities. A principal consideration needed to be floor slipperiness (*the Coefficient of Friction*).

The Expert Panel determined that the slip/fall problem did not end with the selection of suitable flooring material. To maintain its non-slip properties, flooring material had to be protected from wear. The choice of floor treatment (coating) could significantly alter the ability of a flooring material to maintain a suitable coefficient of friction. A non-slippery floor could be rendered slippery if the floor coating was allowed to wear away. As well, a non-slippery floor could be rendered slippery by the application of an improper floor finish.

The Expert Panel determined that floor slipperiness was also a property of ongoing floor maintenance. Clean floors are less slippery than dirty floors. Surface contaminants on a floor not only destroy the floor finish but also act between the sole of a shoe and the floor surface to cause a loss of friction (and resultant slips/falls).

The Expert Panel discovered that floor slipperiness (Coefficient of Friction) could be measured. Some 70 types of slip-test meter (tribometers) are available. However, the Panel discovered that slip-test meters used in Laboratories to measure floor slipperiness although precise cannot be used to evaluate floor slipperiness in the field. Field slip-test (coefficient of friction) data collected using portable slip-test meters does not correlate with Laboratory test data. The presence of contaminants and liquids affect readings. As well, data from tests performed on dry flooring material cannot be used to make judgments about wet flooring material. All these complications raised significant concern in the mind of the Panel about the possibility of achieving a reliable and comparable floor slipperiness measurement.

Notwithstanding these difficulties, the Expert Panel reached a conclusion that in-field testing of floor slipperiness needs to be conducted. The Panel felt that testing inconsistencies between portable slip-test meters could be reduced (perhaps, overcome) if instrumentation was standardized, if common use practices were devised, if field evaluators were trained in the same testing procedure and, if tests were conducted appropriately under these procedures. The Panel felt that, if School Districts commenced such a program of field testing and, if testing results were recorded over time, a *performance database* could be established. School Districts would then be able to pool performance database results and use the information to select better flooring materials, better floor treatment products and better floor maintenance practices thereby lessening the risk of slip/fall incidents/injuries.

The Expert Panel prepared a set of recommendations designed to assist in the reduction of slip/fall incidents. The Expert Panel offers these recommendations to School District decision-makers with the expectation that Districts will find one or more of the suggestions both useful and preventative. The Expert Panel suggests:

1. That the Recommendations Section of the report be distributed to School Districts and School District personnel (maintenance and design). The Expert Panel suggests that the study report Recommendations also be tabled for discussion in upcoming School District meetings (or forums) so that the material elicits discussion.

2. That a follow-up study be conducted in two years' time to determine which recommendations have been adopted in the Districts and which recommendations have realized their preventative potential. The follow-up study would also examine the database records collected in the Districts for correctness and completeness. At that time, the database material could be examined for trends.

Table 9.6. NFSI 101-A Traction Ranges

Wet SCOF Value (μ)	Available Traction	Remediation
μ ≥ 0.60	High traction: lower probability of slipping	Monitor SCOF regularly and maintain cleanliness
0.40 ≤ μ < 0.60	Moderate traction: increased probability of slipping	Monitor SCOF regularly and maintain cleanliness. Consider traction-enhancing products and technologies
μ < 0.40	Minimal available traction: higher probability of slipping	Seek professional intervention. Consider replacing flooring and/or coating with high-traction products

Note: It is important to note that these categories are not indicative of all possible conditions. There are numerous variables that may add to or take from the available traction of any given floor surface. (i.e., type or style of footwear, types and frequency contaminants, pedestrian preoccupation, and so on). Results of dry and wet tests should be viewed independent of each other and not compared.

3. That another set of studies be considered for joint sponsorship; ones broader in scope designed to look at slip/fall incidents in outdoor locations and slip/fall incidents between levels (not just on the same level). The Panel is convinced of the worth of this slip/fall study but "*falls on the same elevation*" and "*falls indoors*" are only part of the overall slip/fall issue.

- Recommendation—A Written Floor Maintenance Program
- Recommendation—Choose a Suitable Flooring Material
- Recommendation—Choose a Suitable Floor Treatment
- Recommendation—Purchase a Slip-Test Meter
- Recommendation—Buy and Use NFSI-Certified Materials
- Recommendation—Investigate All Slip/Fall Incidents
- Recommendation—Use a Standard Investigation Form
- Recommendation—Conduct Education Sessions

RECOMMENDATION—BUY AND USE NFSI-CERTIFIED MATERIALS

Adopt the practice of using product that has been rated as "high traction" by NFSI (USA) and which carries the NFSI certification label.

Why?

Kendzior identifies an initiative being promoted in the United States. According to Kendzior, the National Floor Safety Institute (NFSI) has developed a new standard for product safety. Manufacturers submit their product for two-phase testing.

- Phase 1: Product tested with the NFSI Universal Walkway Tester (UWT) for its wet slip resistance. If it exceeds a SCOF value of 0.6 it is eligible for Phase 2 testing.

- Phase 2: Product placed in a "real work" situation for 30 days; product then cleaned as per manufacturer's instructions; product is re-tested with the UWT. If the SCOF exceeds a value of 0.6, it may be classified as "high traction" and carry the NFSI certification label.

Goodwin (1999) explains the rationale for the NFSI initiative,

> "Since it is impossible to get a tile cleaner than when it was brand new, the tile's COF right out of the box can serve as a value of 100% clean. As the tile becomes dirty, its COF will naturally drop off due to the buildup of dirt and polymerized film. . . . Since the COF of the tile is known, cleaning methods and materials can be evaluated by seeing how close they can come to restoring the tile's original COF."

If School Districts focus on NFSI-certified products, the Districts can expect to have "*safe*" floors (COF >0.5). The School District would use its slip-test meter to confirm the "*safe*" COF value immediately after product application (and, would record the value). The School District would then know what the "*clean coefficient of friction value*" is for the product, as applied.

The School District would then retest the floor at designated intervals.

As the COF value fell (as the floor became "*dirty*"), the School District would be able to define the point where the floor required cleaning (to restore the COF value before the floor became "*unsafe*," e.g., a COF <0.5). Over time, the testing record would allow for forecasting of defensible cleaning regiments (Bonnet et al. 2004).

About the Author

Russell J. Kendzior is one of the nation's leading experts in slip, trip, and fall prevention and through his company, Traction Experts, Inc., has been retained in more than 400 slip, trip, and fall lawsuits. Mr. Kendzior is also the founder and chairman of the board of the National Floor Safety Institute (NFSI), a not-for-profit research and education foundation.

Sought nationwide as a safety consultant, public speaker, and expert witness, Mr. Kendzior is the author of *Slip and Fall Prevention Made Easy* (Government Institutes, Inc., 1999) and is the *OSHA Self-Inspection Checklist.* He has written numerous articles on slip-and-fall accident prevention for such industry publications as *Chain Store Age, ISSA Today, Services Magazine, Occupational Health and Safety,* and *Professional Retail Store Maintenance Magazine (PRSM).* He has been featured in numerous newspaper and magazine articles and has appeared on several nationally televised news programs including *ABC News, Good Morning America,* and *Inside Edition.* He has also been the focus of a documentary on floor safety aired by PBS stations nationwide.

Mr. Kendzior frequently lectures at trade and professional associations, including the National Restaurant Association, the Food Marketing Institute, and the International Sanitary Supply Association. He has also spoken before manufacturing groups representing the floor mat, floor care, and the floor coverings industries. He has been an active participant in the creation of walkway safety standards and is currently the secretary of the ANSI B101 committee on Safety Requirements for Slip, Trip, and Fall Requirements. Mr. Kendzior is a voting member of five American Society for Testing and Materials (ASTM) committees, which develops and publishes a wide range of industry safety standards. He is a past member of the Board of Delegates of the National Safety Council and is a graduate of Bradley University.

You can contact Mr. Kendzior at:

P.O. Box 92628
Southlake, TX 76092
russ@tractionexperts.com

Join the campaign to prevent slips and falls by visiting www.fallsarentfunny.org